T0231722

Efficient Electrical Systems
Design Handbook

Efficient Electrical Systems Design Handbook

by
Albert Thumann, P.E., C.E.M. and
Harry Franz, P.E.

Routledge
Taylor & Francis Group
LONDON AND NEW YORK

Published 2020 by River Publishers

River Publishers

Alsbjergvej 10, 9260 Gistrup, Denmark

www.riverpublishers.com

Distributed exclusively by Routledge

4 Park Square, Milton Park, Abingdon, Oxon OX14 4RN

605 Third Avenue, New York, NY 10017, USA

Library of Congress Cataloging-in-Publication Data

Thumann, Albert.
 Effcient electrical systems design handbook / by Albert Thumann and Harry
Franz.
       p. cm.
    Includes bibliographical references and index.
    ISBN-10: 0-88173-593-0 (alk. paper) -- ISBN-13: 978-8-7702-2278-5 (electronic)
  ISBN-13: 978-1-4398-0300-4 (Taylor & Francis : alk. paper)
  1. Commercial buildings--Electric equipment. 2. Electric power systems.
I. Franz, Harry, 1947- II. Title.

    TK4001.T58 2008
    621.3--dc22

                                                                2008039662

Effcient electrical systems design handbook / by Albert Thumann and Harry Franz.
First published by Fairmont Press in 2009.

Routledge is an imprint of the Taylor & Francis Group, an informa business

10: 0-88173-593-0 (The Fairmont Press, Inc.)
13: 978-1-4398-0300-4 (print)
13: 978-8-7702-2278-5 (online)
13: 978-1-0031-5148-7 (ebook master)

While every effort is made to provide dependable information, the publisher,
authors, and editors cannot be held responsible for any errors or omissions.

*This book is dedicated to my father
who showed me how to install
electrical systems.*

Albert Thumann

Contents

Foreword

The understanding of electrical system design has become increasingly important, not only to the electrical designer, but to safety, plant and project engineers as well. With the advent of high energy costs, plant and project engineers have needed to become more aware of electrical systems. Both safety and energy efficiency will be covered in this text along with practical application problems for industrial and commercial electrical design.

Chapter 1

Electrical Basics

The field of electrical engineering is a large and diverse one. Often included under the general title of electrical engineer are the fields of electronics, semiconductors, computer science, power, lighting and electromagnetics. The focus of this book is on the consulting or plant electrical engineer whose responsibilities include facility power distribution and lighting.

The following chapter provides a brief review of basic concepts that serve as background for the electrical engineer. A thorough understanding of these concepts, while helpful, is not essential to understanding the remainder of this book.

ELECTRICAL UNITS

Table 1-1 and the following text provide definitions of the basic electrical quantities.

Table 1-1. Electrical Quantities in MKS Units

Quantity	Symbol	Definition	Unit
Force	f	push or pull	Newton
Energy	w	ability to do work	joule or watt-second
Power	p	energy / unit of time	watt
Charge	q	integral of current	coulomb
Current	i	rate of flow of charge	ampere
Voltage	v	energy / unit charge	volt
Electric field strength	E	force/unit charge	volt/meter
Magnetic flux density	B	force/unit charge momentum	tesla
Magnetic flux	φ	integral of magnetic weber flux density	weber

"Force"—A force of 1 newton is required to cause a mass of 1 kilogram to change its velocity at a rate of 1 meter per second.

"Energy"—Energy in a system is measured by the amount of work which the system is capable of doing. The joule or watt-second is the energy associated with an electromotive force of 1 volt and the passage of one coulomb of electricity.

"Power"—Power measures the rate at which energy is transferred or transformed. The transformation of 1 joule of energy in 1 second represents an average power of 1 watt.

"Charge"—Charge is a "quantity" of electricity. The coulomb is defined as the charge on 6.24×10^{18} electrons, or as the charge experiencing a force of 1 newton in an electric field of one volt per meter, or as the charge transferred in 1 second by a current of 1 ampere.

"Current"—The current through an area is defined by the electric charge passing through per unit of time. The current is the net rate of flow of positive charges. In a current of 1 ampere, charge is being transferred at the rate of 1 coulomb per second.

"Voltage"—The energy-transfer capability of a flow of electric charge is determined by the potential difference of voltage through which the charge moves. A charge of 1 coulomb receives or delivers an energy of 1 joule in moving through a voltage of 1 volt.

"Electric Field Strength"—Around a charge, a region of influence exists called an "electric field." The electric field strength is defined by the magnitude and direction of the force on a unit positive charge in the field (i.e., force/unit charge).

"Magnetic Flux Density"—Around a moving charge or current exists a region of influence called a "magnetic field." The intensity of the magnetic effect is determined by the magnetic flux density, which is defined by the magnitude and direction of a force exerted on a charge moving in the field with a certain velocity. A force of 1 newton is experienced by a charge of 1 coulomb moving with a velocity of 1 meter per second normal to a magnetic flux density of 1 tesla.

"Magnetic Flux"—Magnetic flux quantity, in webers, is obtained by integrating magnetic flux density over an area.

RESISTANCE

If a battery is connected with a wire to make a complete circuit, a current will flow. (See the schematic representation in Figure 1-1.) The current that flows is observed to be proportional to the applied voltage. The constant that relates the voltage and current is called "resistance." If the symbol v represents volts, the symbol i represents current, and the symbol R represents resistance (measured in ohms-Ω), the relationship can be expressed by the equation:

FORMULA 1-1 $\qquad\qquad\qquad v = Ri$

This expression is called Ohm's Law.

Since voltage is the energy per unit charge and current is the charge per unit time, the basic expression for electrical energy per unit time, or power, is:

FORMULA 1-2 $\qquad\qquad\qquad P = vi = i^2R$

Consequently, resistance is also defined as a measure of the ability of a device to dissipate power (in the form of heat).

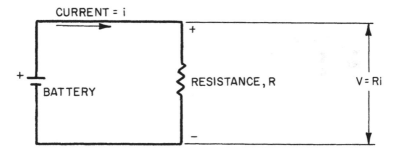

Figure 1-1. Ohm's Law Representation

CAPACITANCE

Now let's connect the battery to two flat plates separated by a small air space between them. (See the schematic representation in Figure 1-2.) When a voltage is applied, it is observed that a positive charge appears on the plate connected to the positive terminal of the battery, and a negative charge appears on the plate connected to the negative terminal. If the battery is disconnected, the charge persists. Such a device that stores charge is called a capacitor.

If a device called a signal generator, which generates an alternating voltage, is installed in place of the battery, the current is observed to be proportional to the rate of change of voltage. The relationship can be expressed by the equation:

FORMULA 1-3 $i = C \, dv/dt$

where C is a constant called "capacitance" (measured in farads) and dv/dt is differential notation representing the rate of change of voltage.

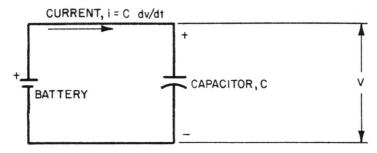

Figure 1-2. Capacitance Law

INDUCTANCE

If the signal generator is placed in a circuit in which a coil of wire is present, it is observed that only a small voltage is required to maintain a steady current. (See the schematic representation in Figure 1-3.) However, to produce a rapidly changing current, a relatively large voltage is required. The voltage is observed to be proportional to the rate of change of the current and can be expressed by the equation:

FORMULA 1-4 $v = L \, di/dt$

where L is a constant called "inductance" (measured in henrys, H) and di/dt is differential notation representing the rate of change of current.

Additionally, when a direct current is removed from an inductor the resulting magnetic field collapses, thereby "inducing" a current in an attempt to maintain the current flow. Consequently, inductance is a measure of the ability of a device to store energy in the form of a magnetic field.

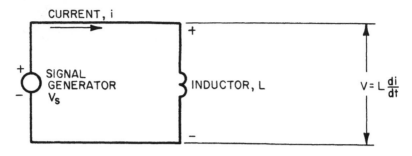

Figure 1-3. Inductance Law

CIRCUIT LAWS

To ease the analysis of complex circuits, circuit laws are utilized. These circuit laws allow voltages and currents to be calculated if only some of the circuit information is known. The two most famous circuit laws are "Kirchoff's Current" and "Voltage Laws." Kirchoff's Current Law states that the sum of the currents flowing into a common point (or node) at any instant is equal to the sum of the currents flowing out.

If current flowing into a node is taken as positive and current flowing out of a node is taken as negative, the summation of all the currents at a node is zero. If the circuit represented by Figure 1-4 is analyzed, Kirchoff's Law would be utilized as follows:

Kirchoff's Voltage Law states that the summation of the voltages measured across all of the components around a loop equals zero. To analyze a circuit using the voltage law, the circuit loop must be traversed in

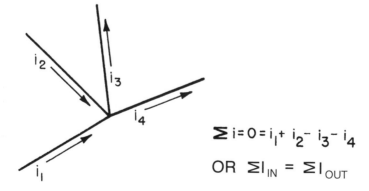

Figure 1-4. Kirchoff's Current Law

one arbitrary direction. If the potential increases when passing through a component in the direction of analysis, the voltage is said to be positive. If the potential decreases, the voltage is said to be negative. Analyzing the circuit of Figure 1-5 gives the following:

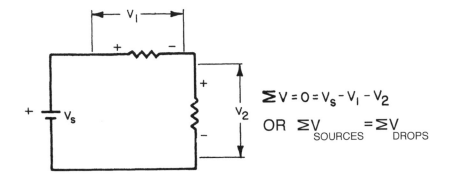

$$\Sigma V = 0 = V_s - V_1 - V_2$$

$$OR \ \Sigma V_{SOURCES} = \Sigma V_{DROPS}$$

Figure 1-5. Kirchoff's Voltage Law

ALTERNATING CURRENT

Although direct current finds uses in some semiconductor circuitry, the primary focus of electrical design is concerned with alternating currents. Among the many advantages of utilizing alternating currents are: the voltage is easily transformed up or down; an alternating current varying at a prescribed frequency provides a dependable time standard for clocks, motors, etc.; and alternating patterns (i.e., sound and light waves) occur in nature and consequently provide the basis for analysis of signal transmission.

An alternating current, or sinusoid, can be represented by the equation below (see Figure 1-6).

FORMULA 1-5 $a = A \cos (\omega t + \alpha)$

where
      a = instantaneous value
      A = amplitude or maximum value
      ω = frequency in radians per second (omega)
      t = time in seconds
      α = phase angle in radians (alpha)

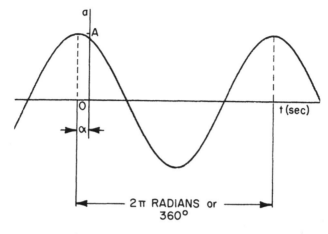

Figure 1-6. Sinusoid Representation

Note that frequency ω is related to frequency (f) in cycles per second, or Hertz (Hz), by:

FORMULA 1-6 $\qquad\qquad\qquad f = \omega/2\pi$

(In the United States, the common power distribution frequency is 60 Hz.)

The phase angle α represents the difference between the reference time t = 0 and the time that the peak amplitude A occurs.

It is convenient to represent Formula 1-5 and Figure 1-6 by a phasor diagram as shown in Figure 1-7.* In this figure, the sinusoid is represented as a complex vector rotating with a frequency ω from an initial phase angle α. Such a concept allows the sinusoid to be represented by complex constants (composed of real and imaginary parts) instead of functions of time and also allows phasors to be added together using the rules of complex algebra.

Using phasor notation for the sinusoid represented by Figure 1-7, we get:

FORMULA 1-7 $\qquad a = b + jc = Ac\,(j\alpha) = A\angle\alpha$

---

*For more information on the phasor concept, see *Circuits, Devices and Systems,* R.J. Smith, John Wiley and Sons, Inc.

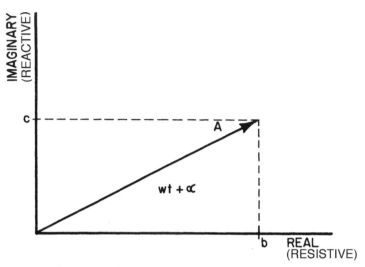

Figure 1-7. Phasor Representation of a Sinusoid

Consequently, the term $A\angle\alpha$ is a simplified representation of Formula 1-5 for a given frequency.

IMPEDANCE

To represent the effect that capacitive and inductive elements have on the current in a circuit, the concept of impedance is introduced.

In a circuit with capacitive and inductive elements present, Ohm's law is modified to be the following, where all symbols are represented by phasors:

FORMULA 1-8 $Z = V/I$

where Z is measured in Ohms.
In a series circuit, Z is defined as:

FORMULA 1-9 $Z = \sqrt{R^2 + X^2} \angle\Theta$

where
      R = resistance
      X = reactance
      Θ = the angle that the voltage phasor leads or lags
          the current phasor

REACTANCE

An inductive element opposes a change in alternating current. We say that the voltage across an inductive element leads by 90° the current through it. Its reactance is given by:

FORMULA 1-10 $\qquad Z_L = V_L / I_L = \omega L \angle 90° = 2\pi f L \angle 90°,$

where $X_L = 2\pi f L$

The voltage across a capacitive element lags by 90° the current through it. Its reactance is given by:

FORMULA 1-11 $\quad Z_C = V_C / I_C = 1/\omega C \angle -90° = 1/2\pi f C \angle -90°,$

where $\qquad\qquad\qquad\qquad X_C = \dfrac{1}{2\pi f C}$

Note that the voltage and current are ill phase across a purely resistive element.

POWER

In a circuit containing resistive elements only, the voltage and current are in phase, and power is calculated by Formula 1-2. When inductive and/ or capacitive elements are present, however, the energy is not dissipated in these elements but is stored and returned to the circuit every half cycle. The current and voltage are not in phase in reactive elements; consequently, Formula 1-12 must be used to calculate power consumption.

FORMULA 1-12 $\quad P = VI \cos \Theta$ watts,
where Θ = angle between V and I

The difference between the power in watts and the "volt-amperes" is the product of a quantity termed the power factor, which is calculated by Formula 1-13.

FORMULA 1-13 $\qquad pf = \cos \Theta = P/VI = (watts/V-A)$

Since in many cases we are interested in the energy transfer capability of an electric current, we often use an effective value for the alternating current. This effective value is found to be the "square root of the mean squared value," or the root mean square (RMS) value. The RMS value produces the same heating effect in a resistance as a direct current of the same ampere value. The RMS value is found to be:

FORMULA 1-14 $I_{RMS} = I_{PEAK} \times 0.707$

FORMULA 1-15 $V_{RMS} = V_{PEAK} \times 0.707$

The RMS values are commonly used when referring to distribution voltages and currents, since this is the value measured by voltmeters and ammeters. Consequently, a rating of 115 volts for an appliance operation is the RMS value. Additionally, RMS values of voltage and current are generally used in Formula 1-12, since average power is generally of primary interest.

SIM 1-1
Using Formula 1-5, write an expression for household voltage (i.e., 115 VAC). Express frequency in terms of Hz and assume the phase angle is zero.

ANSWER
Formula 1-5 is: $a = A \cos(\omega t + \alpha)$. Since 115 VAC is an RMS value, A is 115 volts/0.707 = 163 volts. From Formula 1-6 $\omega = 2\pi f$; therefore, for a 60 Hz distribution frequency the expression is:

$$v = 163 \cos(2\pi 60 t) \text{ volts}$$

THREE-PHASE POWER

Most power is generated and transmitted in three phases in which three wires are utilized, with the voltage in each equal in magnitude but differing in phase by $360°/3 = 120°$ (See Figure 1-8). Three-phase power offers the following advantages over single-phase:
1.   Generators are more efficient.
2.   Motors start and run smoother.
3.   Power is constant rather than fluctuating during the cycle.

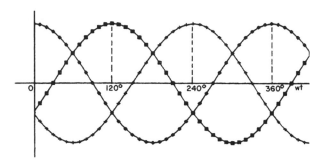

Figure 1-8. Three-phase Voltages

Phasor addition of the currents and voltages of a three-phase power system yield the following expression for power:

FORMULA 1-15   For balanced system
$$\text{watts} = P_{3\phi} = \sqrt{3}\ (V_{L-L})(I_L) \cos \Theta \text{ watts,}$$
where L = line, L–L = line-to-line

FORMULA 1-16 $\text{watts} = P_{3\phi} = 3\ V\phi I\phi \cos \Theta,$
where ϕ = phase

FORMULA 1-17 $\text{VARS} = Q_{3\phi} = 3\ V\phi I\phi \sin \Theta$

FORMULA 1-18 $V\text{–}A = S_{3\phi} = 3\ V\phi I\phi$

TRANSFORMERS

$$\frac{V_{out}}{V_{in}} \ = \ \frac{N_{out}}{N_{in}} \ = \ \frac{I_{in}}{I_{out}} \qquad \text{where N = \# turns}$$

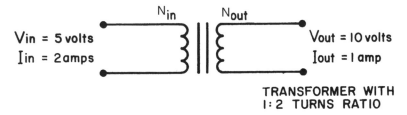

V_{in} = 5 volts
I_{in} = 2 amps

V_{out} = 10 volts
I_{out} = 1 amp

**TRANSFORMER WITH
1:2 TURNS RATIO**

Figure 1-9. Transformer Operation

A transformer is an electrical device which converts alternating voltages and currents from one value to another (either up or down). This is accomplished with approximately equal power transfer (i.e., if the voltage is increased, the current is decreased a proportionate amount). See Figure 1-9.

Transformers contain two separate coils of insulated wire wound on an iron frame. Alternating current flowing through a coil develops a magnetic field that expands and contracts in step with the changes in current. The magnetic field in one coil induces current to flow in the other coil by cutting through the turns of wire.

Transformers are used in many stages in distributing power from the generating station to the user. Power is generally transferred at very high distribution voltages (several thousand volts), since the associated current is relatively low and the distribution losses are much decreased. (Remember that power loss = i^2R and that, for a fixed value of line resistance, the lower the current the much lower the power loss.) Additionally, lower current values allow smaller wires and associated current-carrying equipment to be used. Electrical substations near the point of use, consisting of banks of transformers, are then used to reduce voltages to usable levels.

DISTRIBUTION VOLTAGES

Modern electrical distribution within facilities has tended toward higher voltages, for many of the same reasons as utilities have (i.e., lower costs associated with lower current carrying needs). Consequently, wiring for appliances, outlets, lights, etc. (called branch circuit wiring) has tended to be routed relatively short distances to strategically located transformer load centers, which are then linked to a central distribution panelboard. This approach allows power to be delivered at the required voltage while minimizing long branch circuit runs at low voltage. There are three basic voltage distribution systems that are used today as described below. (Note that more than one of these systems may be present in a facility.)

SINGLE-PHASE

This is a commonly used system in residential and small commercial buildings. See Figure 1-10. 240 volt branch circuits can be used for power

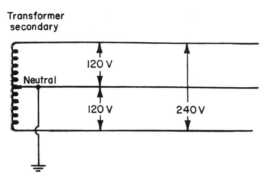

Figure 1-10. Single Phase Distribution

loads such as clothes dryers, electric ranges, etc. 120 volt branch circuits are used for lighting and receptacles.

WYE SYSTEM

This system is designated "wye" because the connection of the transformer secondary coils resembles a "Y." This system provides three-phase and single-phase power at a variety of voltages. See Figure 1-11.

The 120/208 volt configuration is generally available in all except heavy industrial facilities, to some extent. The 120 volt circuits are used for receptacles and lighting, while the 208 volt circuits can be used for motor loads.

The 277/480 volt system is rapidly becoming the system of choice in commercial and industrial facilities because of the advantages mentioned earlier of utilizing higher distribution voltages. 480 volt, three-phase circuits are used for motor loads; 277 volt single-phase circuits are used for fluorescent and HID (high intensity discharge) lighting; 120, 240 or 120/208 volt circuits are available from transformers for receptacles and miscellaneous loads.

In a balanced wye system, the magnitude of line currents equal the phase currents, and the line-to-line voltage is th √3 times the phase voltage.

In a wye system with a fourth neutral wire, and *un*balanced load, the line current magnitude of each particular phase equals the phase current of that phase. Note, however, that the currents from one phase to the other and not necessarily equal, but due to the neutral wire the line-to-neutral voltage remains the same even if the load becomes *un*balanced.

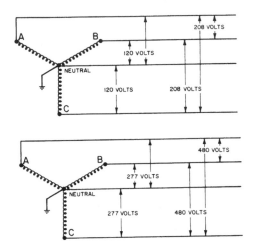

Figure 1-11. Wye Distribution

SIM 1-2

Refer to the second drawing in Figure 1-11. If a given balanced wye load is attached to the source of the system where the wye load has for each of its phases an impedance of (3, 12) with resisstive ohms = 3 and inductive reactive ohms = 12, find the magnitude of the current of each line and the three-phase load watts, vars, v-a and power factor.

Answer

Note (3, 12) rectangular form of the impedance converts to impedance
$Z = 12.4 \angle 76$
V φ load = V line-to-line / (√3) = 480 volts / 1.732 = 277 volts
I φ load = 277 / 12.4 = 22.3 amps
I line = I phase = 22.3 amps
Power factor = p.f. = cos (76 degrees) = 0.24 → 24% lagging.
Note that the power factor lags due to positive reactive ohms load.
$P_{3\phi}$ loads = 3 (277) (22.3) 0.24 = 4.4 kW
$Q_{3\phi}$ loads = 3 (277) (22.3) sin 76 degrees = 18 kVARs
$S_{3\phi}$ loads = 3 (277) (22.3) = 18.53 kVA

DELTA SYSTEMS

The delta-connected secondary system (Figure 1-12) is available with phase-to-phase voltages of 240,480 or 600 volts. This system is used where

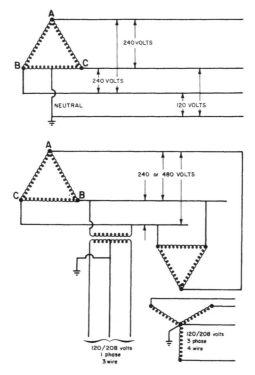

Figure 1-12. Delta Distribution

motor loads represent a large part of the total load (i.e., some industrial facilities). In a typical installation, 480 volt three-phase circuits supply motor loads while lighting and receptacle circuits are supplied by a single or three-phase step-down transformer (as shown in the figure).

A variation on the delta connection is also shown in Figure 1-12. In this system one of the transformer secondary windings is center-tapped to obtain a grounded neutral conductor to the two-phase legs to which it is connected. Motors are supplied at 240 volts three-phase, while 120-volt single-phase circuits are supplied by the neutral conductor and phases B or C.

In a balanced delta system, the line-to-line voltage equals the phase voltage, and the line-current is the √3 times the phase voltage.

SIM 1-3

Refer to the second drawing in Figure 1-12. If a given balanced delta load is attached to the source of the system where the delta load has for each of its phases an impedance of (3, 12) with resisstive ohms =

3 and inductive reactive ohms = 12, find the magnitude of the current of each line and the three-phase load watts, vars, v-a and power factor.

Answer

Note (3, 12) rectangular form of the impedance converts to impedance
$Z = 12.4 \angle 76$

V ϕ load = V line-to-line = 480 volts

I ϕ load = 480/12.4 = 38.7 amps

I line = ($\sqrt{3}$) 38.7 = 67 amps

Power factor = p.f. = cos (76 degrees) = 0.24 → 24% lagging.

Note that the power factor lags due to positive reactive ohms load.

$P_{3\phi}$ loads = 3 (480) (38.7) 0.24 = 13.37 kW

$Q_{3\phi}$ loads = 3 (480) (38.7) sin 76 degrees = 54 kVARs

$S_{3\phi}$ loads = 3 (480) (38.7) = 55.7 kVA

MEASURING ELECTRICAL SYSTEM PERFORMANCE

The ammeter, voltmeter, wattmeter, power factor meter, and foot-candle meter are usually required to do an electrical survey or audit. These instruments are described below.

Ammeter and Voltmeter

To measure electrical currents, ammeters are used. For most audits, alternating currents are measured. Ammeters used in audits are portable and are designed to be easily attached and removed. Ammeters must have relatively low resistance, and are put in series or clamped around.

There are many brands and styles of snap-on ammeters commonly available that can read up to 1000 amperes continuously. This range can be extended to 4000 amperes continuously for some models with an accessory step-down current transformer.

The snap-on ammeters can be either indicating or recording with a printout. After attachment, the recording ammeter can keep recording current variations for as long as a full month on one roll of recording paper. This allows studying current variations in a conductor for extended periods without constant operator attention.

The ammeter supplies a direct measurement of electrical current, which is one of the parameters needed to calculate electrical energy. The second parameter required to calculate energy is voltage, and it is measured by a voltmeter.

A voltmeter measures the difference in electrical potential between two points in an electrical circuit. Voltmeters must have relatively high resistance and are put in parallel.

In series with the probes are the galvanometer and a fixed resistance (which determine the voltage scale). The current through this fixed resistance circuit is then proportional to the voltage, and the galvanometer deflects in proportion to the voltage.

The voltage drops measured in many instances are fairly constant and need only be performed once. If there are appreciable fluctuations, additional readings or the use of a recording voltmeter may be indicated.

Most voltages measured in practice are under 600 volts; there are many portable voltmeter/ammeter clamp-ons available for this and lower ranges.

Several types of electrical meters can read the voltage or current.

Wattmeter and Power Factor Meter

The portable wattmeter can be used to indicate by direct reading the electrical energy in watts. It can also be calculated by measuring voltage, current, and the angle between them (power factor angle).

The basic wattmeter consists of three voltage probes and a snap-on current coil that feeds the wattmeter movement.

The typical operating limits are 300 kilowatts, 650 volts, and 600 amperes. It can be used on both one- and three-phase circuits.

The portable power factor meter is primarily a three-phase instrument. One of its three voltage probes is attached to each conductor phase and a snap-on jaw is placed about one of the phases. By disconnecting the wattmeter circuitry, it will directly read the power factor of the circuit to which it is attached.

It can measure power factor over a range of 1.0 leading to 1.0 lagging, with ampacities up to 1500 amperes at 600 volts. This range covers the large bulk of the applications found in light industry and commerce.

The power factor is a basic parameter whose value must be known to calculate electric energy usage. Diagnostically it is a useful instrument to determine the sources of poor power factor in a facility.

Portable digital kWh and kW demand units are also now available.

Digital units can have read-outs of energy usage in both kWh and kW demand, or in dollars and cents. Instantaneous usage, accumulated usage, projected usage for a particular billing period, alarms when over-target levels are desired for usage, and control-outputs for load-shedding and cycling are possible.

Continuous displays or intermittent alternating displays are available at the touch of a button for any information needed such as the cost of operating a production machine for one shift, one hour, or one week.

Chapter 2

Using the Language of the Electrical Engineer

To design the electrical portions of an industrial plant or a commercial building requires knowledge of power, lighting, and control. The viewpoint in the following chapters is that of an electrical engineer, designing a new facility. This "role playing" experience will enable the reader to gain a better understanding of the elements that go into the design, to deal better with contractors and in-house designers, and to interpret the design of an existing facility.

OBJECTIVES OF ELECTRICAL DESIGN

Three elements usually comprise the basis of electrical design, namely technical proficiency, cost considerations, and overall schedules.

• *Technical Proficiency.* The electrical design should meet the facility's requirements, local and national codes and all safety requirements.

• *Cost Considerations.* Decisions relating to materials of construction, as well as first and operating costs should be analyzed using the principles of life-cycle costing. Electrical engineering and design manpower requirements should be established, monitored and controlled.

• *Schedules.* Schedules should be made for all engineering and construction activities. Delays in engineering or construction activities can be very costly. Monitoring progress, spotting areas of concern, and implementing corrective action are required to ensure an orderly design.

ACTIVITIES OF THE ELECTRICAL ENGINEER/DESIGNER

Throughout this book you will be involved in the design of a hypothetical facility. Since a facility can contain process, power generation, and office areas, you will gain a broad exposure to electrical problems.

It will also be seen that designing the electrical portions of a facility requires both an engineering and a design approach.

Typical Problem

The Ajax Company* is building a plant. The first step for the electrical engineer/designer is to help the client (plant) establish needs. Many clients know what they want, but they need help in defining what has to be done. The design engineer studies all aspects of the client's requirements and establishes the design criteria. The design engineer must consider, for example:

- How to service the loads of the plant. (This includes determining the voltage level to best service the load economically, whether overhead or underground distribution should be used, the type of reliability required, as well as negotiations with the utility company to establish system requirements.)
- Type of lighting system required.
- Type of grounding system required
- Auxiliary systems required.
- Type of equipment required.
- Control considerations.
- Economic considerations.

Once the design criteria have been established the production of the job can begin; designs and specifications to meet the job criteria can be made.

Engineering Activities

Using the above design criteria, the engineering activities include:

- Establishing criteria for One Line Diagram.
- Establishing budgets, schedules and manpower requirements.

---

*The Ajax Company is fictitious, but the principles you will experience are not.

- Writing specifications.
- Inspecting equipment.
- Coordinating activities of design, vendors, subcontractors and client.
- Checking vendor CAD work.
- Performing special studies.
- Preparing estimates.

Design Activities

Design activities are involved in the preparation of the various diagrams required for the project installation. Typical diagrams required for the project are as follows:

- *One-line or Single-line Diagram*
  This diagram (Figure 2-1) is a basic schematic which identifies how power is distributed from the source to the user. This diagram indicates the voltage levels, bus capacities, fuse or breaker ratings, key metering and relaying, and other identification which will aid in describing the electrical distribution. Several one-line diagrams are often needed.

- *Main One-line or Single-line Diagram*
  A main one-line diagram illustrates the distribution of power by the primary switch gear and substations.

- *Motor Control Center (MCC) One-line or Single-line Diagram*
  A main one-line diagram illustrates the distribution of power by the MCC to motors and other loads.

- *Plans*
  A typical plan illustrates by a top view the physical location of equipment. This is to scale and locates equipment in a manner that is similar to that of a map.

- *Power Plans*
  A typical power plan (Figure 2-2) illustrates by a top view the physical location of all motors and other electrical power loads. It is to scale and shows the actual locations of the electrical power users. Conduits and cables are also physically shown on the power plan.

If the project is large, separate diagrams are often used to list the conduit and cable sizes. The location of transformers and power panel boards are also physically shown on the power plan.

- *Lighting Plans*
  A typical lighting plan (Figure 2-5) illustrates by a top view the physical location of all lighting fixtures and other electrical lighting loads. This is drawn to scale and shows the actual locations of the lighting equipment. Conduits and cables for lighting are also physically shown on the lighting plan. The location of lighting transformers and lighting power panel boards are also physically shown on the lighting plan. It is important to note that 120 VAC/240 VAC single-phase receptacles are typically included on the lighting plans rather than the power plans.

- *Instrument Plans*
  A typical instrument plan illustrates by a top view the physical location of all instruments and other instrument-related loads. This is drawn to scale and shows the actual locations of the instruments. Conduits and cables for instruments are also physically shown on the instrument plan. If the project is large, separate diagrams are often used to list more information about instruments and the respective conduit and cable sizes. The location of instrument power transformers, instrument isolation transformers, and instrument power panel boards are also physically shown on the instrument plan. It is important to note that instrument diagrams are not combined with other diagrams on any job, even if the job is very small.

- *Grounding Plans*
  A typical grounding plan illustrates by a top view the physical location of the grounding grid and other grounding items. It is to scale and shows the actual locations of the grounding system, which includes the main grounding loop, grounding branch loops, and location of grounding connections. Details for grounding equipment and columns are shown on a detail as part of the grounding plan or on a separate detail diagram. It is important to note that location of the lightning rods and rod cable downcomers are often shown as part of the grounding plans.

- *Variation/Combination Plans*
  Often on relatively small jobs, plans are combined. Once again, it is important to note that instrument plans are not combined with other plans.

- **Panel Board Schedules**
  Each panel board typically has a schedule which lists the number, location, and power consumed by the panel loads.

- **Power Panel Board Schedules**
  A power panel board schedule lists the number, location, and power consumed by the loads on each branch circuit and the receptacles for three-phase loads. Also, the circuit breaker sizes and the phase to which each is connected are shown.

- **Lighting Panel Board Schedules**
  A lighting panel board schedule (Figure 2-6) lists the number, location, and power consumed by the lights on each branch circuit and the receptacles for single-phase 120 V AC and 240 VAC power loads. Also, the circuit breaker sizes and the panel buss to which each is connected are shown.

- **Instrument Panel Board Schedules**
  An instrument panel board schedule lists the number, location, and power consumed by the instruments on each branch circuit and related instrument power loads. Also, the circuit breaker sizes and the electrical panel buss to which each is connected are shown.

- *Elevations*
  A typical elevation illustrates by a front view, usually for layout, or side view, typically for clarification of details that are used to supplement a plan view.

- *Motor Control Center (MCC) Front View Layout Elevation*
  This shows the actual physical location of motor starters, feeder breakers and feeder fuse-switches and the incoming cable location for the motor control center. The electrical NEMA size and physical space size are shown on the MCC layout. The actual physical arrangement of the starters, breakers, switches, AC drive units, solid-state motor controllers, smart overload relays, power moni-

tors, PLC I/O chassis, and device network communication units are shown.

- **Motor Control Center (MCC) Rear View Elevation**
  This shows the actual physical location of motor control center buss horizontal and vertical, electrical buss bars, and the incoming cable for the motor control center.

- **Lighting Pole Elevation**
  This shows the actual height and aiming of a lighting fixture on a pole, as well as details of the pole base and mounting.

- **Lightning Rod Detail Elevation**
  This shows the actual mounting of the lightning rods, the rod downcomer cables to ground, and the connections to the lightning rods.

- **Grounding Detail Elevation**
  This shows the actual bare or green insulated ground cable upcomer connection to the equipment.

- **Elementary Diagrams**
  The elementary is a type of schematic. A typical elementary diagram (Figure 2-3) is shown for a control system. The elementary may be in ladder logic form, as shown in the figure. Typical control devices, such as pushbuttons, limit switches, level switches, and pressure switches (all of which are inputs) are used to energize relays, motor control coils, and solenoids values (all of which are the outputs). Note that the elementary diagram indicates how the system operates but not the physical attributes of each element.

- **Interconnection Diagrams**
  A typical interconnection diagram (Figure 2-4) is based on the elementary diagram. Terminal numbers and point-to-point connections are shown in their relative location on the interconnection diagram for each device. A schedule that lists this information is often used in place of this diagram. It is important to note that the interconnection diagram shows connections from one device to another, as well as the details of connections at each device.

- **Connection Diagrams**
  A typical connection diagram is also based on the elementary diagram. Terminals number and point-to-point are shown in their relative location on the connection diagram for single device. A schedule that lists this information is often used in place of this diagram. It is important to note that a connection diagram shows only connections to a device.

- **Details and Miscellaneous Drawings**
  Details are often used to supplement plans. The details may indicate more information about a special item, more precise item locations, a larger or different view of an item, or any further clarifications necessary.

- **Critical Path Diagram**
  The critical path diagram is a type of time schedule (Figure 2-7). It shows which items are dependent on the completion of other items and which items are independent from the completion of other items. Items in series are dependent on the completion of other items. The previous items in series must be done before the next item in series. On the other hand, items in parallel or shunt shown beside each other may be independently completed.

DESIGN ACTIVITY HOURS

When dealing with outside contractors and evaluating the scope of a project, it is useful to have an understanding of how work time estimates are made.

Table 2-1 summarizes the CAD work, plus the hours required. The hours for each item depend on the amount of detail shown and the type of firm making the design. For instance, an architectural firm designing a commercial lighting project may simply choose the fixture type and design a layout. A designer in an engineering firm would probably show in addition the circuits to each fixture and a lighting panelboard schedule. The more detail, the more hours would be required. This table should serve only as a hypothetical example. The practices in each firm and past job performances will be the prevalent factors in determining the CAD details and the estimated hours.

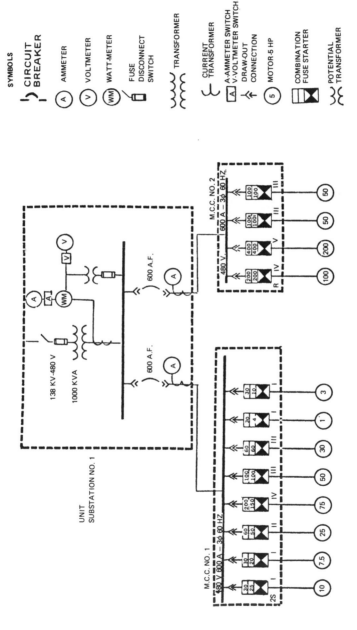

Figure 2-1. Typical One-line Diagram

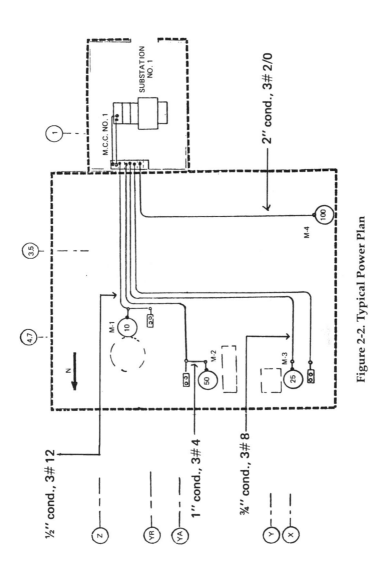

Figure 2-2. Typical Power Plan

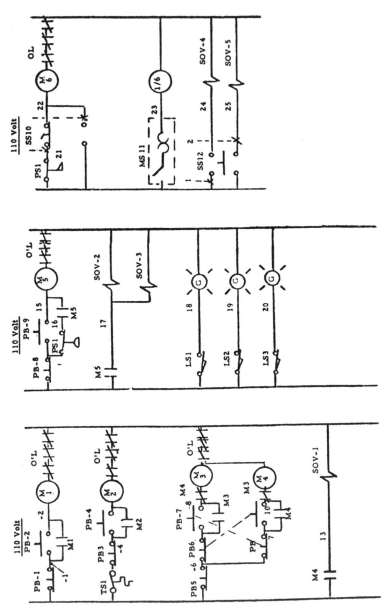

Figure 2-3. Typical Elementary Diagram

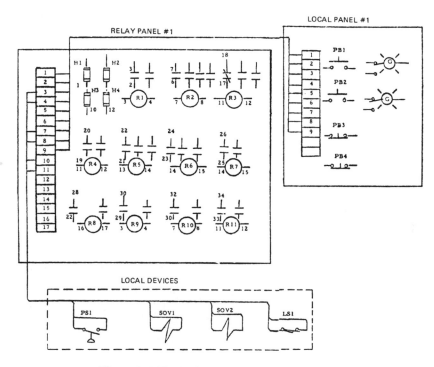

Figure 2-4. Typical Interconnection Diagram

Table 2-1. Typical Job Requirements

ITEM	SCALE	HOURS TO DESIGN *(including calculations)*
Lighting	1/8"=1'	30-50
Power	1/4"=1'	50-75
Lighting Schedules	None	10
Conduits & Cable Lists	None	10
Elementary Wiring	None	75-100
*One Lines	None	75-100
Interconnection	None	30-50
Grounding	1"=100' (Depends on Plot Plan)	30-40
Miscellaneous Details	Depends on Detail	50-100 Depends on Detail

A unit substation and fifty motors can usually be fitted on a full-size, one-line diagram.

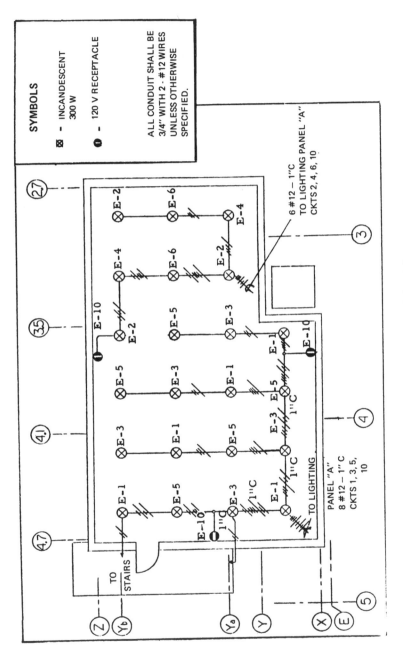

Figure 2-5. Typical Lighting Plan

LIGHTING PANEL "A"

CIRCUIT NO.	SERVICE	NO. OF OUTLETS	WATTS	NEUTRAL A B C	WATTS	NO. OF OUTLETS	SERVICE	CIRCUIT NO.
1	WAREHOUSE	5	1500		900	3	WAREHOUSE	2
3	WAREHOUSE	5	1500		600	2	WAREHOUSE	4
5	WAREHOUSE	4	1500		600	2	WAREHOUSE	6
7	H&V UNITS	-	320		990	11	WAREHOUSE STAIRS	8
9	BLANK	-	-		600	3	RECEPTACLE	10
11	BLANK		-		1200		SPARE	12
13								14
15								16
17								18
19								20
21								22
23								24

12 CIRCUIT PANEL
20 AMPERE BREAKER
120/208 V – 3 PHASE
4 WIRE

PANEL LOADING	
PHASE	WATTS
A	3710
B	2700
C	2100
CONNECTED LOAD	8510
SPARES	1200
TOTAL	9710

Figure 2-6. Typical Lighting Panel

In the following pages you will experience job situations. Each simulation experience will be denoted by SIM. The answer will be written below the problem. Cover the answer so that you can play the game.

SIM 2-1

Estimate the number of power and lighting layouts for the following details.

Client: Ajax manufacturing plant "A"
              Basement—50' × 200'
              Operating Floor—50' × 200'

Answer

From Table 2-1

Power Plan—1/4"=1'

Thus 50' × 200' will fit on a 12" × 50" layout.

Since standard size is 30" × 42", half of the operating and basement floor can fit on one layout. Two layouts are required.

Lighting Plan—1/8"=1'

$$\frac{50}{8} \times \frac{200}{8} = 6'' \times 25''$$

One layout will be sufficient (30" × 42").

SIM 2-2

Estimate the number of one-line diagrams for the plant of SIM 2-1.

Given: 28 motors on two motor control centers with an estimated load of 800 kVA. Assume one substation will feed load.

Answer

Based on the above load, one unit substation and the associated motor control centers will fit on one diagram.

SIM 2.3

Estimate the number of elementary and interconnection diagrams for SIM 2-2. Each motor needs a separate stop-start control scheme. Control schemes should be provided so that 14 solenoid valves can be activated. All relays, pushbuttons, etc., are located on a local panel. Assume 100 elementary lines per diagram. Allow 2 spaces between each scheme.

Answer

Estimating elementary and control schemes is a very difficult task. Usually at the beginning of the project it is difficult to get an exact description of the control.

Assume stop-start scheme with two lines per scheme; thus 28 × 2 = 56 lines. Assume 2 spaces between each motor.

Scheme: 28 × 2 = 56 lines

Assume 2 lines per each solenoid scheme, 28 × 2 = 56 lines

Since more than 100 lines are required, two diagrams are estimated.

In estimating the interconnection diagrams, assume an interconnection diagram is needed for each motor control center and local panel.

SIM 2-4

Compile a list with estimated hours for problems SIM 2-1 through 2-3.

Answer

DESCRIPTION	ESTIMATED HOURS
One Line Diagram	75
Power Plan	50
Power Plan	50
Lighting Diagram—Basement Operating	30
Grounding Drawing	30
Conduit and Cable Schedule—Assume 2*	20
Lighting Schedule—Assume 2	20
Elementary Diagram	75
Elementary Diagram	75
Interconnection MCC No. 1	30
Interconnection MCC No. 2	30
Interconnection Local Devices	_30_
TOTAL HOURS	515

*One for basement and one for operating floor.

ENGINEERING ACTIVITY HOURS

It is more difficult to evaluate the engineering activities at the beginning of the project, because many of the activities are involved with intangibles such as coordination.

From the details of the design, an estimate can be made of the number of requisitions which are needed to purchase equipment.

- To write a specification using a previous one as a guide may take from 10 to 15 hours.

- To write a completely new specification may take 60 hours.

- Compiling material requisitions for quotation and purchase, evaluating sellers' quotes, and checking vendor prints could vary from 40 to 100 hours per requisition. The hours required depend on the complexity of the equipment being purchased.

Many times engineering activities may be estimated as a percentage of design hours. Coordination and general engineering activities may be from 10-20% for small projects below 5000 hours and 15-30% for larger ones that need more coordination. Specifications, requisitions and special studies should be added in separately.

SIM 2-5

Estimate the engineering hours for the design of 500 hours as indicated in SIM 2-4. Assume no special studies; past specifications are available and minimum coordination time is required.

Answer

Substation Specification	10
Requisition and Vendor Items	40
Motor Control Centers (2)	
Specifications	10
Requisition and Vendor Items	40
Coordination—10% (500)	_50_
	150 hours

Notice that the engineering activities on an industrial project are usually only a fraction of the total time required for design.

TYPES OF DESIGN ORGANIZATIONS

The design section of the facilities department of a plant or that of an outside consulting firm is usually organized either by departments or by tasks. The two organizations widely encountered are:

- *Department Oriented.* In the department oriented firm, each group (electrical, structural, civil, architectural, HVAC, piping, and plumbing) is separate from the other. The department head usually has people grouped together. All information usually channels down from the department head.

- *Task Force.* Each department head chooses people for a project. These people move out of the department and operate as a team. Each individual usually gains responsibility since the departmental chain of command has disappeared.

  In the project or task force approach, a project engineer is assigned to the job to help coordinate the various disciplines.

KNOW THY VENDOR

Manufacturers provide many services in order to sell their products. Good relations with manufacturers can aid in electrical design.

Typical services include:

- Computer analysis for determining the most economical lighting system.
- Technical brochures.
- Cost information.
- Proven expertise in their field.

Since most of these services are free, good communications with vendors is extremely important. Remember, vendors are offering these services as a means of selling their product. Be careful not to get attached to one vendor and to look objectively at all information offered.

ELECTRICAL SCHEDULE

Always Last

Electrical design cannot proceed without the motor horsepowers. For example, the heating and ventilating group must size their fans, the mechanical group must select their pumps, and the architectural group must select the automatic roll-up doors before all motor horsepower can be given to the electrical engineer. The electrical engineer is vulnerable to any changes from the other departments, since such changes will probably affect the electric load.

Other aspects which affect design are firm equipment locations. This input is required to design the power and lighting. A description of operation or a logic diagram is needed before the elementary and interconnections can be designed. Thus the electrical design is usually the last to be finished on a project. Because of this, the electrical engineer is always under pressure to complete the design.

Critical Path

In many cases it is the electrical group which determines the critical path. Remember that the delivery of electrical equipment, such as switchgear or high voltage bus duct, may take up to a year to fabricate. Thus, one of the first activities the electrical engineer should do is purchase equipment. This means that many times an estimate of electrical loads must be used to purchase equipment in order to meet the schedules.

A typical critical path schedule is illustrated in Figure 2-7.

JOB SIMULATION SUMMARY PROBLEM

SIM 2-6

Background

At the end of several chapters, a job simulation summary problem will be given. In these problems you will play the role of an electrical en-

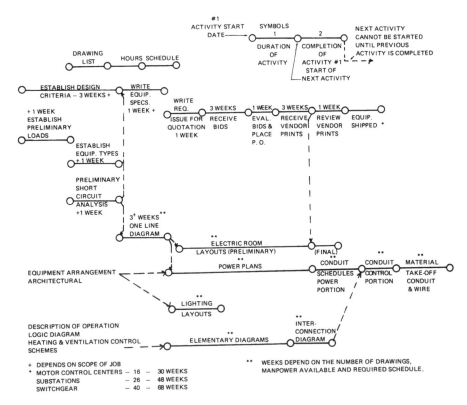

Figure 2-7. Critical Path of Electrical Activities for Plant Design

gineer working on Process Plant 2 of a grass roots (new) project for the Ajax Corporation. The plant is comprised of two identical modules. The response to each of these problems will be needed in order to complete the subsequent chapters.

The first task will be to identify the electrical loads from an equipment list which has been given to the engineer by the plant. From this list, indicate which loads require electric power and from which group you would expect to receive the information.

For example:

    A—Architectural
    M—Mechanical
    H—Heating and Ventilating

Equipment List—Module #1
Note: Module #2 is identical.

D-1	Tank #1
AG-1	Agitator for Tank #1
H-2	Heat Exchanger
CF-3	Centrifuge
FP-4	Feed Pump
TP-5	Transfer Pump
CTP-6	Cooling Tower Feed Pump
D-7	Water Chest
V-8	Vessel #1
CT-9	Cooling Tower
HF-10	H&V Supply Fan
HF-11	H&V Exhaust Fan
UH-12	Unit Heater
BC-13	Brine Compressor
A-14	Air Operated Motor
T-15	Turbine #1
C-16	Conveyor
H-17	Hoist
ES-18	Exhaust Stack
SC-19	Self-cleaning Strainer
VS-20	Vacuum Separator
PO-21	Pneumatic Oscillator
RD-22	Roll-up Door

The plant has also indicated that one unit substation will handle the load and that each module should be fed from a motor control center.

The area of the Process Plant 2 is as follows:

Basement   20 ft × 400 ft
Operating Floor   20 ft × 400 ft

From the above information prepare a list and submit the cost to do this project. Assume the project is complex (use higher hours) and that the unit rate includes overhead, profit and fees.

Analysis

Once it is known what equipment requires a motor, it is necessary to obtain the electrical loads from each discipline and compile a motor information file. A completed motor information file for each module would look as follows:

EQUIP. NO.	NAME	HP	TYPE	DESIGN DISCIPLINE
AG-1	Agitator	60		M
CF-3	Centrifuge	100	Reversing	M
FP-4	Feed Pump	30		M
TP-5	Transfer Pump	10		M
CTP-6	Cooling Tower Feed Pump	25		M
CT-9	Cooling Tower	20	2.speed	M
HF-10	H&V Supply Fan	40		H
HF-11	H&V Exhaust Fan	20		H
UH-12	Unit Heater	1/6		H
BC-13	Brine Compressor	50		M
C-16	Conveyor	20	Reversing	M
H-17	Hoist	5	Local starter by vendor	M
SC-19	Self-cleaning Strainer	3/4		M
RD-22	Roll-up Door	1/8	Local starter by vendor	A

Next, prepare an information file to estimate the hours required to produce the diagrams and specifications.

Typical Estimate

Design

DESCRIPTION	ESTIMATED HOURS
One Line Diagram	100
Power Plan—Basement	75
Power Plan—Operating Floor	75
Lighting Diagram—Basement and Operating Floor	50
Grounding Drawing	40
Conduit & Cable Schedule (2)	20
Lighting Schedule (2)	20
Elementary Diagram	100
Elementary Diagram	100
Interconnection MCC No. 1 and Local Device	50
Interconnection MCC No. 2 and Local Device	_50_
	680 Hrs

Engineering

Substation Specification	15
— Requisition and Vendor Items	100
Motor Control Centers	15
— Specifications, Requisitions, Vendor Items	100
Engineering Coordination—20% (680)	_130_
Total	360

TOTAL HOURS	1040—say 1100 Hrs
TOTAL COST	Hrs × (cost/hr)

The estimate is probably on the high side, since the two modules are identical and the time to do the second module would be less. Also it was assumed that the job is complex.

Estimating is an "art" rather than an exact science.

Chapter 3

Equipment Selection Considerations

The electrical engineer is responsible for distributing power from the service entrance to its varied destinations throughout the facility. This chapter introduces some of the equipment that the engineer must size and specify when designing a safe and efficient power distribution system. Motor control center layout as well as circuit breaker selection considerations are presented in this chapter.

ELECTRICAL EQUIPMENT

Electrical equipment commonly specified is as follows:
*   *Switchgear Breakers*—used to distribute power and provide overcurrent protection for high voltage applications.

*   *Unit Substation*—used to step down voltage. Consists of a high voltage disconnect switch, transformer, and low-voltage breakers. Typical transformer sizes are 300 kVA, 500 kVA, 750 kVA, 1000 kVA, 2000 kVA, 2500 kVA and 3000 kVA.

*   *Motor Control/Center (MCC)*—a structure which houses starters and circuit breakers, or fuses for motor control, and various other control devices. It consists of the following:
    (1) Thermal overload relays which guard against motor overloads;
    (2) Fuse disconnect switches or breakers which protect the cable and motor and can be used as a disconnecting means;
    (3) Contactors (relays) whose contacts are capable of opening and closing the power source to the motor.
    (4) AC drive units, which control AC motor speed.
    (5) Solid-state motor controllers, which start, stop and reverse motors.

(6) Smart overload relays, which protect equipment from high currents.

(7) Power monitors, which check equipment load status.

(8) PLC I/O chassis, for programmable logic controlleers.

(9) Network communication, which sends data between equipment.

• *Panelboard/Switchboard Breakers*—used to distribute power and provide overcurrent protection to motor control centers, lighting, receptacles, and miscellaneous power circuitry within a building.

MOTOR CONTROL CENTER BREAKERS AND FUSES

It should be noted that substation breakers are different from switchgear breakers, motor control center breakers, and panelboard breakers. Switchgear breakers may be of the "vacuum type," whereas substation breakers may be of the "magnetic air circuit type," and motor control center and panelboard breakers may be of the "molded case" type.

One of the most important ratings for a circuit breaker is its ability to interrupt a short circuit (also called a fault). When a short circuit occurs, many thousand amperes of peak (or asymmetrical) current can flow which can "fuse" the contacts of a circuit breaker closed, thereby preventing it from operating to prevent fires, explosions, etc. The interrupting capacity (see Table 3-1) rating of a circuit breaker determines its ability to "clear" a fault of a certain magnitude. Typical low-voltage breakers (less than 600 volts) breakers are illustrated in Figure 3-1.

Note that low-voltage circuit breakers often have two trip ranges. The first, or "thermal," protection is adjusted to prevent the wiring from overheating when overloaded for a period of time. The second, "instantaneous trip," is often adjustable and is used to quickly interrupt a circuit, should the high levels of current associated with a short circuit occur.

Table 3-1 summarizes breaker and starter sizes for motor control centers. Table 3-2 summarizes dual-element fuse and switch sizes for motor control centers. The fuse and breaker sizes indicated in these tables are based on vendor's data. As long as the values are below specified values listed in the National Electrical Code and coordinate with the motor, the selection is satisfactory.

Table 3-1. Combination Breaker-starter Ratings

MOTOR HP	NEMA STARTER SIZE	BREAKER TRIP* Amps	BREAKER FRAME** Amps	1 SPACE FACTOR = 14" VERTICAL HEIGHT†	BREAKER TYPE	TRIP RANGE	ASYM. AMPS
1	1	15	100	14"	FA	15-100	15,000
1½	1	15	100	14"	JA	70-225	20,000
2	1	15	100	14"	KA	70-225	25,000
3	1	15	100	14"	LA	125-400	35,000
5	1	15	100	14"			
7½	1	30	100	14"			
10	1	40	100	14"			
15	2	50	100	14"			
20	2	50	100	14"			
25	2	50	100	14"			
30	3	70	100	28"			
40	3	100	100	28"			
50	3	100	100	28"			
60	4	125	225	42"			
75	4	150	225	42"			
100	4	200	225	42"			
125	5	225	225	70"			
150	5	300	400	70"			
200	5	350	400	70"			

* Check chosen vendor for specific recommendations.

** Minimum size — check short-circuit rating.

† Based on MCC Vendor's data for FVNR (Full Voltage Non-Reversing Starters). Check chosen vendor for specific details.

NOTE: Space factors vary according to manufacturer.

TYPE	DESCRIPTION
Molded Case Breakers	Current passes through a bimetallic strip, which generates heat during an overload. The instantaneous response occurs when the solenoid activates the plunger to break the current path.
High IC Breakers	Breakers are molded from glass fiber or epoxy, rather than standard bakelite materials. Higher fault interruption promotes poor current limiting. These breakers are 50% more expensive than molded case types.
Fuse Circuit Breakers and Limiters	These breakers combine the convenience of breakers with high fault current-limiting characteristics of fuses.
Current Limiting Circuit Breakers	These breakers operate on the principle that the higher the fault current, the stronger the magnetic force induced, which results in a more rapid interruption. This is accomplished by a reverse current loop, which creates a magnetic force causing the conductors to repel. Current-limiting circuit breakers have the advantages of fused breakers and are available with solid state tripping.
Motor Circuit Protectors	These breakers are standard types without thermal element and must be used with overload relays.
Solid State Breakers	These breakers are highly adjustable. Features include easy adjustments to ampere rating, long time delay band, short time delay pickup, short time delay band, instantaneous pickup, ground fault pickup and ground fault time delay. Solid state breakers use current sensors to provide a signal to the trip unit proportional to the magnitude of the current.

Figure 3-1. Typical Breaker Types

There are several types of fuses commonly used. Each type is characterized by its time to isolate the fault, interrupting rating, and current limiting property. Fuse types include standard fuses, dual element fuses, limiters, fast-acting fuses, and semiconductor fuses. Typical fuses are illustrated in Figure 3-2.

Table 3-2a. Full-load Current in Amperes, Direct-current Motors
The following values of full-load currents* are for motors running at base speed.

Horsepower	*Armature Voltage Rating**					
	90 Volts	120 Volts	180 Volts	240 Volts	500 Volts	550 Volts
1/4	4.0	3.1	2.0	1.6	—	—
1/3	5.2	4.1	2.6	2.0	—	—
1/2	6.8	5.4	3.4	2.7	—	—
3/4	9.6	7.6	4.8	3.8	—	—
1	12.2	9.5	6.1	4.7	—	—
1-1/2	—	13.2	8.3	6.6	—	—
2	—	17	10.8	8.5	—	—
3	—	25	16	12.2	—	—
5	—	40	27	20	—	—
7-1/2	—	58	—	29	13.6	12.2
10	—	76	—	38	18	16
15	—	—	—	55	27	24
20	—	—	—	72	34	31
25	—	—	—	89	43	38
30	—	—	—	106	51	46
40	—	—	—	140	67	61
50	—	—	—	173	83	75
60	—	—	—	206	99	90
75	—	—	—	255	123	111
100	—	—	—	341	164	148
125	—	—	—	425	205	185
150	—	—	—	506	246	222
200	—	—	—	675	330	294

*These are average dc quantities.

Reprinted with permission from NFPA 70-2008, National Electrical Code®, Copyright© 2007, National Fire Protection Association, Quincy, Massachusetts 02269. This reprinted material is not the complete and official position of the N2FPA on the referenced subject, which is represented only by the standard in its entirety.
National Electrical Code® and NEC® are Registered Trademarks of the National Fire Protection Association, Inc., Quincy, MA.

Advantages of Fuses Over Breakers
The advantages of fuses over breakers are:
- Higher interrupting ratings (200,000 amps dual-element fuses)
- Current limiting action)will limit the short circuit current downstream of fuse)
- Lower cost
- Less affected by corrosive atmosphere
- Less affected by moisture
- Less affected by dust
- Faster clearing

Table 3-2b. Full-load Currents in Amperes, Single-phase Alternating-current Motors

The following values of full-load currents are for motors running at usual speeds and motors with normal torque characteristics. The voltages listed are rated motor voltages. The currents listed shall be permitted for system voltage ranges of 110 to 120 and 220 to 240 volts.

Horsepower	115 Volts	200 Volts	208 Volts	230 Volts
1/6	4.4	2.5	2.4	2.2
1/4	5.8	3.3	3.2	2.9
1/3	7.2	4.1	4.0	3.6
1/2	9.8	5.6	5.4	4.9
3/4	13.8	7.9	7.6	6.9
1	16	9.2	8.8	8.0
1-1/2	20	11.5	11.0	10
2	24	13.8	13.2	12
3	34	19.6	18.7	17
5	56	32.2	30.8	28
7-1/2	80	46.0	44.0	40
10	100	57.5	55.0	50

Advantages of Breakers Over Fuses

The advantages of breakers over fuses are:

- Resetable
- Electrically operated breakers can be remotely operated.
- Adjustable characteristics

SIM 3-1

Indicate the starter size for 10,30 and 100 HP motors.

Answer (from Table 3-1)

10 HP	Size I
30 HP	Size III
100 HP	Size IV

SIM 3-2

The short-circuit current available at a motor control center is 15,000 amperes asymmetrical. Indicate the frame (continuous rating) and trip (current at which breaker will open) sizes for 7-1/2, 30, 60 and 100 HP motors.

Table 3-2c. Full-load Current, Two-phase Alternating-current Motors (4-Wire)

The following values of full-load current are for motors running at usual speeds for belted motors and motors with normal torque characteristics. Current in the common conductor of a 2-phase, 3-wire system will be 1.41 times the value given. The voltages listed are rated motor voltages. The currents listed shall be permitted for system voltage ranges of 110 to 120, 220 to 240, 440 to 480, and 550 to 600 volts.

	Induction-type Squirrel Cage and Wound Rotor (Amperes)				
Horsepower	115 Volts	230 Volts	460 Volts	575 Volts	2300 Volts
1/2	4.0	20	1.0	0.8	—
3/4	4.8	2.4	1.2	1.0	—
1	6.4	3.2	1.6	1.3	—
1-1/2	9.0	4.5	2.3	1.8	—
2	11.8	5.9	3.0	2.4	—
3	—	8.3	4.2	3.3	—
5	—	13.2	6.6	5.3	—
7-1/2	—	19	9.0	8.0	—
10	—	24	12	10	—
15	—	36	18	14	—
20	—	47	23	19	—
25	—	59	29	24	—
30	—	69	35	28	—
40	—	90	45	36	—
50	—	113	56	45	—
60	—	133	67	53	14
75	—	166	83	66	18
100	—	218	109	87	23
125	—	270	135	108	28
150	—	312	156	125	32
200	—	416	208	167	43

Answer (from Table 3-1)

Horsepower	Trip Size	Frame Size
7-1/2	30	100
30	70	100
60	125	225
100	200	225

SIM 3-3

The short-circuit current available at a motor control center is 25,000 amperes asymmetrical.

Repeat SIM 3-2.

Table 3-2d. Full-load Current, Three-phase Alternating-current Motors

The following values of full-load currents are typical for motors running at usual speeds for belted motors and motors with normal torque characteristics.

The voltages listed are rated motor voltages. The currents listed shall be permitted for system voltage ranges of 110 to 120, 220 to 240, 440 to 480, and 550 to 600 volts.

Horse-power	Induction-Type Squirrel Cage and Wound Rotor (Amperes)							Synchronous-Type Unity Power Factor* (Amperes)			
	115 Volts	200 Volts	208 Volts	230 Volts	460 Volts	575 Volts	2300 Volts	230 Volts	460 Volts	575 Volts	2300 Volts
1/2	4.4	2.5	2.4	2.2	1.1	0.9	—	—	—	—	—
3/4	6.4	3.7	3.5	3.2	1.6	1.3	—	—	—	—	—
1	8.4	4.8	4.6	4.2	2.1	1.7	—	—	—	—	—
1-1/2	12.0	6.9	6.6	6.0	3.0	2.4	—	—	—	—	—
2	13.6	7.8	7.5	6.8	3.4	2.7	—	—	—	—	—
3	—	11.0	10.6	9.6	4.8	3.9	—	—	—	—	—
5	—	17.5	16.7	15.2	7.6	6.1	—	—	—	—	—
7-1/2	—	25.3	24.2	22	11	9	—	—	—	—	—
10	—	32.2	30.8	28	14	11	—	—	—	—	—
15	—	48.3	46.2	42	21	17	—	—	—	—	—
20	—	62.1	59.4	54	27	22	—	—	—	—	—
25	—	78.2	74.8	68	34	27	—	53	26	21	—
30	—	92	88	80	40	32	—	63	32	26	—
40	—	120	114	104	52	41	—	83	41	33	—
50	—	150	143	130	65	52	—	104	52	42	—
60	—	177	169	154	77	62	16	123	61	49	12
75	—	221	211	192	96	77	20	155	78	62	15
100	—	285	273	248	124	99	26	202	101	81	20
125	—	359	343	312	156	125	31	253	126	101	25
150	—	414	396	360	180	144	37	302	151	121	30
200	—	552	528	480	240	192	49	400	201	161	40
250	—	—	—	—	302	242	60	—	—	—	—
300	—	—	—	—	361	289	72	—	—	—	—
350	—	—	—	—	414	336	83	—	—	—	—
400	—	—	—	—	477	382	95	—	—	—	—
450	—	—	—	—	515	412	103	—	—	—	—
500	—	—	—	—	590	472	118	—	—	—	—

*For 90 and 80 percent power factor, the figures shall be multiplied by 1.1 and 1.25, respectively.

Table 3-2e. Combination Fuse-switch Ampere Ratings

MOTOR HP	460 V FLA	FUSE *	SWITCH	MCC** SPACE	SPACE FACTOR
1	1.8	4	30	14"	1 Sf
1½	2.6	5	30	14"`	"
2	3.4	8	30	14"	"
3	4.8	10	30	14"	"
5	7.6	15	30	14"	"
7½	11	20	30	14"	"
10	14	25	30	14"	"
15	21	30	30	14"	"
20	27	40	60	14"	"
25	34	50	60	14"	"
30	40	60	60	14"	"
40	52	80	100	28"	2 Sf
50	65	100	100	28"	"
60	77	125	200	42"	3 Sf
75	96	150	200	42"	"
100	124	200	200	42"	"
125	156	250	400	70"	5 Sf
150	180	300	400	70"	"
200	240	400	400	70"	"

* Based on dual-element fuses.
** Based on MCC vendor's data for FVNR starters.
　Check chosen vendor for specific details.

Answer (from Table 3-1)

Horsepower	Trip Size	Frame Size	
7-1/2	70	225	Min. Size
30	70	225	Breaker
60	125	225	KA—25,000
100	200	225	

Note: For large short-circuit currents, breakers are impractical as indicated above. Either the short-circuit current should be decreased, or fuses should be used instead of breakers.

TYPE	*DESCRIPTION*
Dual-Element Fuses	These fuses have dual-element construction giving them a long time delay characteristic in the overload region and current-limiting in the short circuit region (200,000 amperes interrupting capacity).
Fast-Acting Fuses (Non-Time Delay)	These non-time delay fuses are single-element construction and have a very high degree of current limitation in the short circuit region. A major difference with dual-element fuses is that the fast-acting fuses do not have a long-time delay characteristic in the overload region (200,000 amperes interrupting capacity).
Low Peak Dual-Element Fuses	Similar to dual-element fuses, these fuses are considerably more current-limiting in the short circuit region.
Limiters	These fuses provide only short circuit current-limiting protection and are not designed to provide overload protection.
Semiconductor Fuses	These special fuses have an extremely high degree of current limitation and are used to protect power superconductors.

Figure 3-2f. Typical Fuse Types

SIM 3-4

Indicate the fuse and switch sizes for the following: 3, 10, 50, 75 HP motors.

Answer (from Table 3-2e)

Horsepower	*Fuse Size*	*Switch Size*
3	10	30
10	25	30
50	100	100
75	150	200

MOTOR CONTROL CENTER LAYOUTS

Figure 3-3 shows a typical outline for an MCC. Dimensions vary between vendors, and the space allocated for wiring depends on whether cables enter from top or bottom and whether terminal blocks are required in upper or lower section.

Usually terminal blocks are located in the individual starter cubicle, and the top or bottom is used only for wiring between sections.

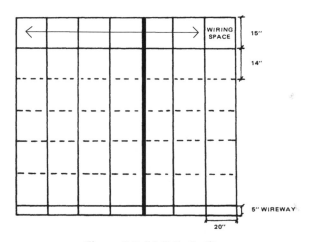

Figure 3-3. M.C.C. Outline
(Consult Vendor for Specific Dimensions)

Referring to Figure 3-3, a vertical section is a structure with an overall height of 90″, a width of 20″ and a depth of nominally 13″-20″. It includes a horizontal feeder bus at the top and a vertical bus bar to accept the plug-in motor and starter units. A unit is the motor starter-disconnect module that fits into the vertical section and is covered by a door. Thus, the motor control center consists of vertical sections bolted together and connected with a common horizontal feeder bus bar. The working height is the available space in any section for motor control center units.

Layout

When designing a motor control center keep in mind the following:

• Larger units should be placed near the bottom of a section for easier maintenance.

- Place the various required units in as many sections as necessary to accommodate them.

- All units fit into the vertical section merely by moving the unit support brackets to fit the structure. Filler plates can be used for leftover spaces.

- The commonly used sizes that are used in conjunction with Figure 3-1 are summarized in Figure 3-4. The sizes shown in Figure 3-4 are based on combination fuse starters.

MCC One-line Diagram

Typical space factors used for an MCC are shown in Figure 3-4. Typical symbols used for an MCC one-line diagram are illustrated in Figure 3-5.

MOTORS

Squirrel Cage Induction Motors

AC induction motors have a rotating magnetic field in the stator that induces a field in the rotor causing it to rotate without any brushes or commutators. (Invented by Nicola Tesla to replace the more complicated and higher maintenance DC motors that have commutators and brushes.) AC induction motors have pairs of poles per each phase to create the rotating electromagnetic field in the stator which mechanically does not move, but induces a rotating field into the rotor.The speed of the rotor depends on the frequency into the motor and the number of poles which come in pairs.

$$\text{Frequency} = \frac{\text{\# of poles x speed}}{120} \qquad \text{(Formula 3-1a)}$$

$$\text{SPEED RPM } (mech.) = \frac{\text{Frequency (hertz electrical)} \times 60 \text{ sec/minute}}{\text{(number of pairs of poles per phase)}} \qquad \text{(Formula 3-1b)}$$

Full-voltage Non-reversing AC Induction Motor (FVNR)

Basically this starts across the line and either runs or stops. The single-phase FVNR motor requires two power leads, and the three-phase versions require six power leads.

FVNR = FULL VOLTAGE NON-REVERSE
FVR = FULL VOLTAGE REVERSE
TSCP = 2 SPEED SINGLE WINDING
TSSW = 2 SPEED SEPARATE WINDING

(NOTE: ALL SPACE FACTORS 20" WIDE)
FILLER PLATE: 1 SPACE FACTOR (14" HIGH) THRU 5 SPACE FACTORS (70" HIGH)

ONE SPACE FACTOR 14" HIGH:

NEMA SIZE ONE FVNR
NEMA SIZE TWO FVNR
30 AMP FUSED SWITCH
60 AMP FUSED SWITCH

TWO SPACE FACTORS (28" HIGH):

NEMA SIZE THREE FVNR
NEMA SIZE ONE FVR
NEMA SIZE TWO FVR
NEMA SIZE ONE TSCP FVNR
NEMA SIZE THO TSCP FVNR
200 AMP FUSED SWITCH

THREE SPACE FACTORS (42" HIGH):

NEMA SIZE FOUR FVNR
NEMA SIZE THREE FVNR
NEMA SIZE ONE FVNR
NEMA SIZE TWO FVNR
200 AMP FUSED SWITCH

FOUR SPACE FACTORS (56" HIGH):

NEMA SIZE FOUR FVR
NEMA SIZE THREE TSCP FVR
NEMA SIZE THREE TSSW FVNR

FIVE SPACE FACTORS (70" HIGH):

NEMA SIZE FIVE FVNR
NEMA SIZE FIVE FVR
NEMA SIZE FOUR TSCP FVNR
NEMA SIZE FIVE TSCP FVNR
NEMA SIZE FOUR TSCP FVR
NEMA SIZE FIVE TSCP FVR
NEMA SIZE FOUR TSSW FVNR
NEMA SIZE FIVE TSSW FVNR

Note: The minimum bus for a motor control center is 600 amperes. Initial sizing is usually based on 400 to 500 HP per motor control center. This is usually adequate. A detailed load check should be made when the design is firm.

Figure 3-4. Figure list: typical space factors for MCC

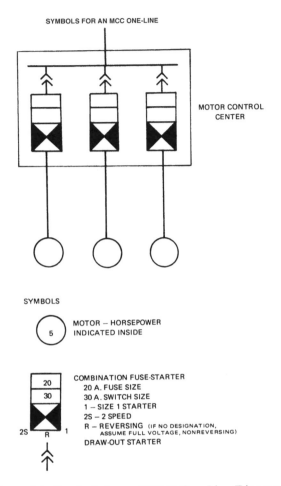

SYMBOLS FOR AN MCC ONE-LINE

MOTOR CONTROL
CENTER

SYMBOLS

MOTOR – HORSEPOWER
INDICATED INSIDE

COMBINATION FUSE-STARTER
20 A. FUSE SIZE
30 A. SWITCH SIZE
1 – SIZE 1 STARTER
2S – 2 SPEED
R – REVERSING (IF NO DESIGNATION,
ASSUME FULL VOLTAGE, NONREVERSING)
DRAW-OUT STARTER

Figure 3-5. Symbols for an M.C.C. One-Line Diagram

Full-voltage Reversing AC Induction Motor (FVR)

This starts across the line and either runs forward or backward, or stops. The single-phase motor requires two power leads, and the three-phase version requires three power leads. The reversing action is accomplished with the controls switching the power leads to the motor. A single-phase AC induction motor will reverse if its two power leads are swapped. The three-phase version will reverse its direction if only two of the three power leads are exchanged. Exchanging just any two of the three power leads to the three-phase motor will change the phase-sequence and cause reversal. Note again that the exchange of leads is accomplished through the control circuit which is shown in a later chapter.

Full-voltage Two-speed Consequent Pole AC Induction Motor (TSCP)
    This starts across the line and either runs fast or slow, or stops. The single-phase motor requires three power leads, and the three-phase version requires four or more power leads. The speed action is accomplished by the controls using certain power leads to the motor. A single-phase AC induction motor will run one speed if the leads to the full set of poles of the motor are chosen. The motor will run the other speed if a lead is chosen so that only part of the set of poles is used.

Full-voltage Two-speed Separate Winding AC Induction Motor (TSSW)
    This starts across the line and either runs fast or slow, or stops. The single-phase motor requires four power leads, and the three-phase version requires six power leads. The speed action is accomplished by the controls using either one set of windings and poles for the slow speed, or the other set of windings and poles for the fast speed. The three-phase version that uses six power leads will use either one set of three power leads for slow, or the other set of three power leads for fast. The motor is like two or three motors built into one, with a complete set of power leads for each speed.

    Thus, if the frequency is fixed, the effective number of motor poles should be changed to change the speed. This can be accomplished by the manner in which the windings are connected. Three-phase two-speed motors require six power leads.
    DC motors are used where speed control is essential. The speed of a DC motor is changed by varying the field voltage through a rheostat, or through higher efficiency solid state controls. A DC motor requires two power wires to the armature and two smaller cables for the field.
    Synchronous Motors are used when constant speed operation is essential. Synchronous motors are sometimes cheaper in the large horsepower categories when slow speed operation is required. Synchronous motors also are considered for power factor correction. A .8 P.F. synchronous motor will supply corrective kVARs to the system. A synchronous motor requires AC for power and DC for the field. Since many synchronous motors are self-excited, only the power cables are

required to the motor.

Energy Efficient versions of induction motors are available which can offer significant energy savings depending on the application. Additionally, these motors offer significant improvement in power factor. See Figures 3-6 and 3-7.

Variable Speed Drives utilizing inverters are being used to change the frequency of power (and thereby the speed) supplied to standard induction motors. Since some motor applications, primarily for pumps and fans, can utilize reduced speed operation for much of the operating time, significant energy savings are possible. (Note, that a 50% reduction in speed yields an 88% reduction in power required for a fan.)

MOTOR HORSEPOWER

The standard power rating of a motor is referred to as a horsepower. In order to relate the motor horsepower to a kilowatt (kW), multiply the horsepower by .746 (conversion factor) and divide by the motor efficiency.

$$kW = \frac{HP \times .746}{\eta} \qquad \text{(Formula 3-2)}$$

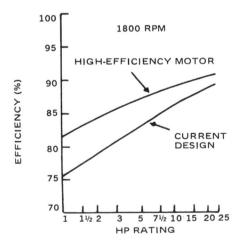

Figure 3-6. Efficiency vs Horsepower Rating (Dripproof Motors)

HP = Motor Horsepower
  η = Efficiency of Motor

Motor efficiencies and power factors vary with load. Typical values are shown in Table 3-3. Values are based on totally enclosed fan-cooled motors (TEFC) running at 1800 RPM "T" frame.

Table 3-3.

HP RANGE	*3-30*	*40-100*
η% at		
1/2 Load	83.3	89.2
3/4 Load	85.8	90.7
Full Load	86.2	90.9
P.F. at		
1/2 Load	70.1	79.2
3/4 Load	79.2	85.4
Full Load	83.5	87.4

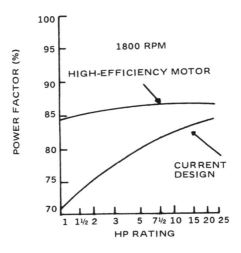

Figure 3-7. Power Factor vs Horsepower Rating (Dripproof Motors)

MOTOR VOLTAGES

For fractional motors, typically 115 volt, single-phase is used. These motors are usually fed from a lighting or panel and do not appear on the MCC one-line diagram. A local starter consisting of a switch and over-load element is all that is usually required.

Motors 1/2 HP to 250 HP are usually fed from a 480-volt, 3-phase, motor control center or equivalent.

Motors 300 HP and above are usually fed at 2300 or 4160 volts. The reason for this is mostly economics, i.e., price of motor, starter, cable and transformer.

SIM 3-5a

Determine the number of poles for a 3600 RPM motor (60 cycle service).

Answer

$$\text{No. of poles} = \frac{\text{Frequency} \times 120}{\text{Speed}} = \frac{60 \times 120}{3600} = 2$$

SIM 3-5b

A single-phase induction motor is rated 115 VAC, 60 HZ, 4 pole, and 1/2 HP. Find the RPM rated speed.

Answer

$$\text{SPEED (RPM)} = (60 \text{ HZ} \times 60 \text{ sec/min})/(2 \text{ pairs}) =$$
$$3600 \text{ RPM}/2 = 1800 \text{ RPM}$$

SIM 3-5c

A three-phase induction motor is rated 115 VAC, 50 HZ, 6 pole, and 10 HP. Find the RPM rated speed.

Answer

SPEED(RPM) = (50 HZ × 60 sec/min)/(3 pairs/3phases) = 3000 RPM

SIM 3-6

Next to each motor indicate the probable voltage rating and whether it is a single- or three-phase motor. Motor HP are: 1/4, 25, 150, and 400.

Answer

MOTOR HP	VOLTAGE	NUMBER OF PHASES
1/4	110 V	1φ
25	460 V	3φ
150	460 V	3φ
500	2300 or 4.16 KV	3φ

SUMMARY

Circuit breakers and fuses are used to protect wiring and equipment from fault currents. When a group of motors is being supplied with power, the circuit breakers and/or fuses are mounted in a motor control center.

Motor efficiency and power factor are dependent on size and load. Energy saving motors offer significant improvement in both efficiency and power factor.

JOB SIMULATION—SUMMARY PROBLEM
SIM 3-7
Background

(a) From the motor list established in SIM 2-6 at the end of Chapter 2, indicate the rated voltage of each motor and if it is a single- or three-phase.

(b) The electrical engineer has been asked by the client to issue a requisition for direct purchase of the motor control centers. Assume a 30 Amp switch in each MCC to take care of lighting loads and fractional horsepower motors. Make a sketch of the proposed layout and an MCC one-line diagram. Assume each module is fed from a separate MCC. Use the horsepower and data provided. Typical forms are illustrated.

Forms

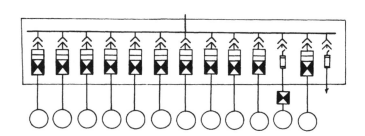

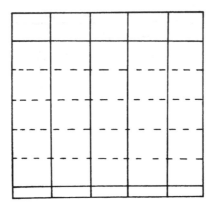

Analysis

(a) The client reviewed the motor list and expected to see the following:

Motor List—Module 1

MOTOR NO.	DESCRIPTION	HP	VOLTAGE	PHASE
AG-1	Agitator Motor	60	460	3
CF-3	Centrifuge Motor	100	460	3
FP-4	Feed Pump Motor	30	460	3
TP-5	Transfer Pump Motor	10	460	3
CTP-6	Cooling Tower Feed Pump Motor	25	460	3
CT-9	Cooling Tower Motor	20	460	3
HF-10	H&V Supply Fan Motor	40	460	3
HF-11	H&V Exhaust Fan Motor	20	460	3
UH-12	Unit Heater Motor	1/6	110	1
BC-13	Brine Compressor Motor	50	460	3
C-16	Conveyor Motor	20	460	3
H-17	Hoist Motor	5	460	3
SC-19	Self-Cleaning Strainer Motor	3/4	460	3
RD-22	Roll-Up Door Motor	1/8	110	1

(b) A typical response the client would expect to SIM 3-7 would look as follows:

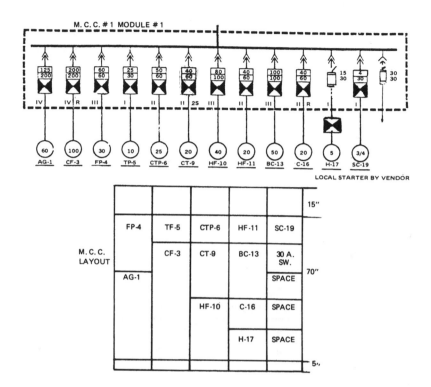

Chapter 4

Analyzing Power Distribution Systems

The electrical engineer initiates one-line diagrams for new facilities and interprets one-line diagrams of existing facilities. This chapter illustrates some of the simple concepts involved in establishing a one-line diagram, determining system reliability, and determining breaker interrupting capacities.

Three substation quotes will be analyzed, and a vendor will be recommended at the end of the chapter.

THE POWER TRIANGLE

The total power requirement of a load is made up of two components, namely the resistive part and the reactive part. The resistive portion of a load cannot be added directly to the reactive component, since it is essentially ninety degrees *out* of phase with the other. The pure resistive power is known as the watt, while the reactive power is referred to as the reactive volt amperes. To compute the total volt ampere load it is necessary to analyze the power triangle indicated below:

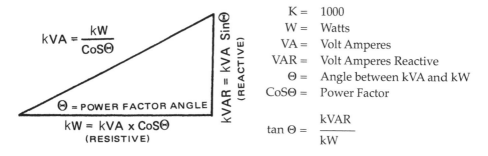

$$kVA = \frac{kW}{Cos\Theta}$$

$$\Theta = \text{POWER FACTOR ANGLE}$$

$$kW = kVA \times Cos\Theta$$
(RESISTIVE)

$$kVAR = kVA\ Sin\Theta$$
(REACTIVE)

K =	1000
W =	Watts
VA =	Volt Amperes
VAR =	Volt Amperes Reactive
Θ =	Angle between kVA and kW
$Cos\Theta$ =	Power Factor

$$\tan \Theta = \frac{kVAR}{kW}$$

POWER FACTOR CORRECTIONS EXAMPLE

ORIGINAL PLANT LOAD: 1000 KVA @ pf =70% Lag

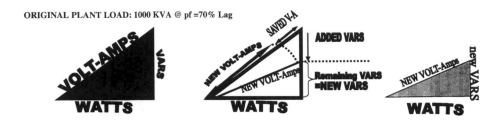

FIND THE ADDED VARS REQUIRED TO CORRECT THE POWER FACTOR TO 87% LAG.

FIRST FIND THE POWER TRIANGLE VALUES FOR THE GIVEN ORIGINAL PLANT LOAD:

1000 KVA @ 70% LAG $\rightarrow$ pf = .7 , $\theta = \cos^{-1}(.7) = 45.5$ degrees ,

original KW=1000 (.7) = 700 KW ,   original KVARS=1000 sin 45.5=1000 (.713)= 713 KVARS

NEXT FIND REQUIRED VALUES FOR THE NEW POWER TRIANGLE:

TO CORRECT TO 87% THE NEW POWER FACTOR,

The NEW ANGLE IS = $\theta = \cos^{-1}(.87) = 29.5$ degrees, and the SIN 29.5 degrees = .492

NOTE THAT THE WATTS, KW ARE THE SAME IN BOTH THE NEW AND ORIGINAL POWER TRIANGLE.

NEW WATTS= ORIGINAL WATTS= 700 KW ,

NEW KVA= (700 KW / .87) = 805 KVA ,   NEW VARS = 805 KVA (.492) = 396 KVARS

CHECK: $\left[\sqrt{new\ 700\ KW^2\ +\ new\ 396\ KVARS^2}\right] = new\ 804\ KVA,$ which checks

NOTE: 1000 KVA- 804 KVA= 196 KVA SAVED

FINALLY FIND THE REQUIRED ADDED VARS TO GIVE THE NEW VARS:

ORIGNAL VARS OF ORIGINAL POWER TRIANGLE = 713 KVARS
NEW VARS OF NEW POWER TRIANGLE= 396 KVARS
THE DIFFERENCE IS (713 KVARS - 396 KVARS) = 317 KVARS WHICH ARE REQUIRED TO BE ADDED

IF GIVEN f= 60 HZ, VOLTAGE= 13.8KV AC, for example, to obtain ADDED "C" REQUIRED: VARS=(V^2 / X) ,

$$X_C = \left[\frac{(13.8\ KV)^2}{317\ KVARS}\right] = 600\ (ohms)\ ,\quad X_C = \left\{\frac{1}{2\Pi f C}\right\} \longrightarrow C = \left\{\frac{1}{(2\Pi f)\ X_C}\right\} = \frac{1}{(377)(600)} = 4.421\ \mu F$$

IMPORTANCE OF POWER FACTOR

Transformer size is based on kVA. The closer Θ equals $0°$ or power factor approaches unity, the smaller the kVA. Many times utility companies have a power factor clause in their contract with the customer. The statement usually causes the customer to pay an additional power rate if the power factor of the plant deviates substantially from unity. The utility company wishes to maximize the efficiency of their transformers and associated equipment. Similarly, a facility wishes to maximize the efficiency of its distribution equipment by minimizing current levels with a high power factor.

Many larger institutions are billed for electrical energy use in *kVA*-hours rather than *kilowatt*-hours, which is done for residences.

The loads of many institutions and industry are usually more inductive loads that include motors and transformers. This causes the apparent power kVA to be larger than the "real" energy, which is in kilowatts. The equipment must be sized larger to provide the same amount of energy. The utility bills the consumer for this.

To counterbalance this situation, power factor correcting capacitors are usually added to correct the power factor close to the ideal value of 100%, usually at about 95%. The same amount of "real" energy watts is provided more effectively.

Power factor correction can often be done for an entire site; however, power factor correction is also often done at various individual locations and even at individual pieces of equipment. The utility bill will be reduced for a more ideal power factor and the existing equipment may be used to carry more kVA load. This is often done on project expansions to "free up" the load of equipment, rather than purchase and replace existing equipment with larger units, transformers, for example.

The power factor correcting capacitor units have an initial cost, but the "payback" is often worth it. Also if it is desired to obtain a power factor that is very close to the 100% ideal power, say 98%, at least some of the capacitors should have an automatic switching feature. These automatic switching capacitors switch in and out as the load demand changes, and are often more expensive.

Power Factor Correction

The problem facing the electrical engineer is to determine the power factor of the plant and to install equipment such as capacitor banks or synchronous motors so that the overall power factor will meet the utility company's objectives.

Capacitor banks lower the total reactive kVAR by the value of the capacitors installed.

SIM 4-1

A total motor horsepower load of 854 is made up of motors ranging from 40-100 horsepower. Calculate the connected kVA. Refer to Table 3-3 for efficiency and power factor values.

Answer

$$kVA = \frac{HP \times .746}{Motor\ Eff. \times Motor\ PF}$$

From Table 3-3 at full load
    Eff. = .909 and PF = .87

$$kVA = \frac{854 \times .746}{.909 \times .87} = 806$$

SIM 4-2

As an initial approximation for sizing a transformer assume that a horsepower equals a kVA. Compare answer with SIM 4-1.

Answer

    Total HP = 854
    HP = kVA = 854
    Problem answer = 806

For an initial estimate, equating horsepower to kVA is done in industry. This accuracy is in general good enough, since motor horsepowers will probably change before the design is finished. The load at which the motor is operating is not established at the beginning of a project, and this approach usually gives a conservative answer.

SIM 4-3

It is desired to operate the plant of SIM 4-1 at a power factor of .95. What approximate capacitor bank is required?

Answer

From SIM 4-1 the plant is operating at a power factor of .87. The power factor of .87 corresponds to an angle of 29°.

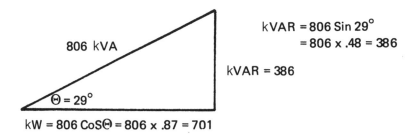

$$kVAR = 806 \sin 29°$$
$$= 806 \times .48 = 386$$

$$kVAR = 386$$

806 kVA

$$\Theta = 29°$$

$$kW = 806 \cos\Theta = 806 \times .87 = 701$$

A power factor of .95 is required.

$$\cos\Theta = .95$$
$$\Theta = 18°$$

The kVAR of 386 needs to be reduced by adding capacitors.

Remember kW does not change with different power factors, but kVA does.

Thus, the desired power triangle would look as follows:

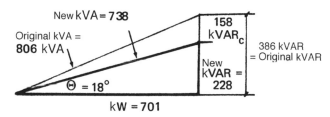

New kVA = 738

Original kVA = 806 kVA

$$\Theta = 18°$$

158 kVAR$_c$

386 kVAR = Original kVAR

New kVAR = 228

kW = 701

$$\cos\Theta = .95$$
$$\Theta = 18° \qquad \sin 18° = .31$$

$$kVA_c = \frac{701}{\cos\Theta} = \frac{701}{.95} = 738$$

Note: Power factor correction reduces total kVA.

$$kVAR_c = 738 \sin 18° = 738 \times .31 = 228$$
$$\text{Capacitance Bank} = 386 - 228$$
$$\cong 158 \; kVAR_c$$

SIM 4-4

A plant original load is 5 MVA @ pf = 80% lag. Find the (capacitive) MVARS$_c$ added to correct the power factor to 95% lag, which is often the practical desired value for the power factor.

Answer

Using the original power triangle 5 MVA × 0.8 original power factor = 4 MW original. Again using the original power triangle and noting that a power triangle is a right triangle, $(5^2 - 4^2)^{1/2}$ = 3 MVARS original.

Correcting the power factor does not change the wattage; hence, the new MW = original MW = 4 MW. The new power triangle has a desired power factor of 0.95, lag so the new MVA = (new MW/0.95, the new power factor) = 4.21 MVA new.

Noting again that a power triangle is a right triangle, (4.21 MVA new^2 – 4 MW^2 new)$^{1/2}$ = 1.31 MVARS new.

The original MVARS – new MVARS gives the added MVARS. Original 3 MVARS – 1.31 new MVARS = 1.69 added capacitive MVARS$_c$.

SIM 4-5

A plant original load is 200 kVA @ pf = 90% lag; find the (capacitive) kVARS$_c$ added to correct the power factor to 100% in-phase, the ideal power factor.

Answer

Using the original power triangle 200 kVA × 0.9 original power factor = 180 kW original. Using the original power triangle and noting that a power triangle is a right triangle, (200 kVA^2 – 180 kW^2)$^{1/2}$ = 87.2 kVARS original.

The new power triangle has a desired power factor of 100% in-phase, so the new power triangle has zero kVARS. Note that its pf = cos zero = 1.

The original kVARS – new kVARS gives the added kVARS. Original 87.2 kVARS – 0.0 new kVARS = 87.2 added capacitive kVARS$_c$.

SIM 4-6

Assume that all motors in SIM 4-1 are not running at the same time. The diversity factor, which takes into account the cycle time, is assumed from previous plant experience to be 1.1. Indicate the minimum transformer size.

Answer

$$kVA_{Min.} = \frac{kVA_+}{Diversity\ Factor} = \frac{806}{1.1} = 832$$

Many times in industry the transformer capacity is simply based on the sum of the motor horsepowers plus an additional factor to take into account growth. The conservative sizing approach may not be too exact, but it does allow for normal changes in design and growth capacity. Remember that electrical loads seldom shrink.

WHERE TO LOCATE CAPACITORS

As indicated, the primary purpose of capacitors is to reduce the reactive power levels. Additional benefits are derived by capacitor location. Figure 4-1 indicates typical capacitor locations. Maximum benefit of capacitors is derived by locating them as closely as possible to the load. At this location, its kilovars are confined to the smallest possible segment, decreasing the load current. This, in turn, will substantially reduce power losses of the system. Power losses are proportional to the square of the current. When power losses are reduced, voltage at the motor increases; thus, motor performance also increases.

Locations C1A, C1 Band C1C of Figure 4-1 indicate three different arrangements at the load. Note that in all three locations, extra switches are not required, since the capacitor is either switched with the motor starter or the breaker before the starter. Case C1A is recommended for new installations, since the maximum benefit is derived and the size of the motor thermal protector is reduced. In Case C1B, as in Case C1A, the capacitor is energized only when the motor is in operation. Case C1B is recommended in cases where the installation is existing, since the thermal protector does not need to be re-sized. In position C1C, the capacitor is permanently connected to the circuit, but it does not require a separate switch, since it can be disconnected by the breaker before the starter.

It should be noted that the rating of the capacitor should *not* be greater than the no-load magnetizing kVAR of the motor. If this condition exists, damaging overvoltage or transient torques can occur. This is why most motor manufacturers specify maximum capacitor ratings to be applied to specific motors.

The next preference for capacitor locations, as illustrated by Figure

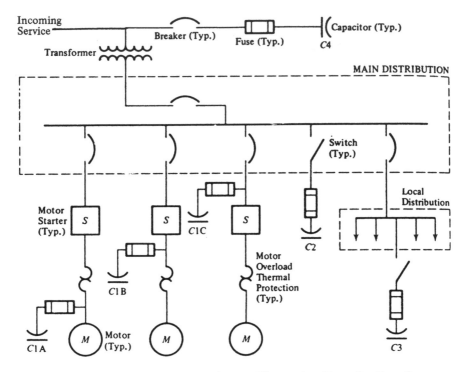

Figure 4-1. Power Distribution Diagram Illustrating Capacitor Locations

4-2, is at locations C2 and C3. In these locations, a breaker or switch will be required. Location C4 requires a high voltage breaker. The advantage of locating capacitors at power centers or feeders is that they can be grouped together. When several motors are running intermittently, the capacitors are permitted to be on line all the time, reducing the total reactive power regardless of load.

POWER FLOW CONCEPT

Power flowing is analogous to water flowing in a pipe. To supply several small water users, a large pipe services the plant at a high pressure. Several branches from the main pipe service various loads. Pressure reducing stations lower the main pressure to meet the requirements of each user. Similarly, a large feeder at a high voltage services a plant. Through switchgear breakers, the main feeder is distributed into smaller feeders.

The switchgear breakers serve as a protector for each of the smaller feeders. Transformers are used to lower the voltage to the nominal value needed by the user.

HOW TO CREATE A ONE-LINE DIAGRAM

An overall one-line diagram indicates where loads are located and how they are fed.

- The first step is to establish loads and their locations by communicating with the various engineers. (Remember that initial design is based on your best estimate.)

- The next step is to determine the incoming voltage level based on available voltages from the utility company and the distribution voltage within the plant.
    (a) For single buildings and small complexes without heavy equipment loads, incoming voltage levels may be 208 or 480 volts. For small industrial plants to 10,000 kVA, voltage levels may be 2300, 4160, 6900 or 13.8 kV.
    (b) For medium plants, 10,000 kVA to 20,000 kVA voltage levels may be 13.8 kV.
    (c) For large plants above 20,000 kVA, 13.8 kV or 33 kV are typical values.

    The advantages of the higher voltage levels are:
    (a) Feeders and feeder breakers can handle greater loads. (Certain loads are more economical.)
    (b) For feeders which service distant loads, voltage drops are not as noticeable on the higher voltage system.

- The third step is to establish equipment types, sizes and ratings.

- The last step is to determine the system reliability required. The type of process and plant requirements are the deciding factors. The number of feeds and the number of transformers determine the degree of reliability of a system.

The three commonly used primary distribution systems for industrial plants are simple radial, primary selective, and secondary selective systems.

- The *simple radial system* is the most economical. As Figure 4-2 indicates, it is composed of one feed and one transformer.

- The *primary selective system* is composed of two feeds and two primary transformer disconnect switches. See Figure 4-3.

- The *secondary selective system* is the most reliable and the most expensive. As Figure 4-4 indicates, it is composed of two complete substations joined by a tie breaker.

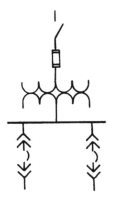

Figure 4-2.
Simple Radial System

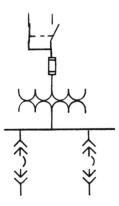

Figure 4-3.
Primary Selective System

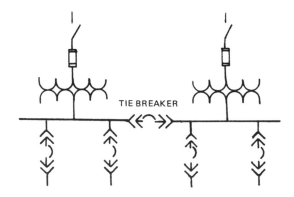

Figure 4-4. Secondary Selective System

SIM 4-7

Develop the one-line diagram for the Ajax Plant. The various steps for developing a one-line diagram are illustrated in SIM 4-7 through SIM 4-11.

Step 1: Establish loads and their locations.

	Approximate Load (kVA)
Administration Building	210
Machine Shop	340
Warehouse	185
Boiler House	675
Process Unit No. 1	890
Process Unit No. 2	765

Locate loads on Figure 4-5.

Answer

See Figure 4-6.

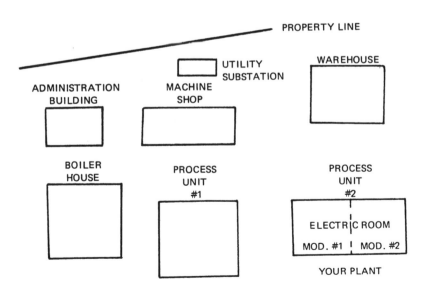

Figure 4-5. Ajax Plant Layout

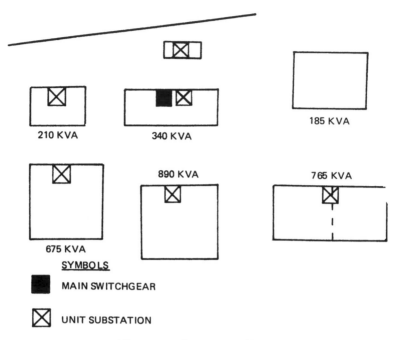

210 KVA

340 KVA

185 KVA

675 KVA

890 KVA

765 KVA

SYMBOLS

MAIN SWITCHGEAR

UNIT SUBSTATION

Figure 4-6. Answers to SIM 4-7

SIM 4-8

Step 2: Determine distribution voltage to substations for SIM 4-5. Assume plant capacity may triple in the future and the utility primary voltage is 115 kV.

Answer

Total load is 3065, which represents the small plant category. Since the plant load may triple in the future, 13.8 kV would be recommended as it meets the present load with future expansion.

SIM 4.9

Step 3: Establish equipment sizes for SIM 4-7. Substations should be located as closely as possible to the load center. If loads are small (below 300 kVA) and are located near another load center, consideration should be given to combining it with other small loads.
Use standard size transformers illustrated in Table 4-1, and size at least 1.25 times given load.

	Load
Administration Building	210
Feed from same Transformer	
Machine Shop	340
Warehouse	185
Boiler House	675
Process Unit No. 1	890
Process Unit No. 2	765
Total Load	3065

Locate substations and switchgear on Figure 4-5.

Step 4: Comment on the system reliability recommended, assuming a noncritical process.

Answer	Load	Transformer
	1.25	Size
Administration Building	262	300 kVA
Feed from same Transformer		
Machine Shop	425	
Warehouse	230	750 kVA
Boiler House	845 ⎫	1000 kVA
Process Unit No. 1	1112 ⎭	1500 kVA
Process Unit No. 2	956	1000 kVA

The main feeder from the utility transformer should be sized to meet the total load of 3065, plus capacity for the future. It becomes impractical to run this size feeder to each substation. Thus the main switchgear is provided to distribute power to the various substations.

The substations and switchgear are located on Figure 4-6.

Unless the process is critical, a simple radial system is commonly used.

SIM 4-10

Make a one-line diagram for SIM 4-7. Assume that up to 500 kVA can be put on each motor control center. Provide two switchgear breakers, one feeding the process substations and the second feeding the utility and auxiliary areas.

Answer

See Figure 4-7.

GENERAL TIPS FOR DISTRIBUTION SYSTEMS

• Always size unit substations with growth capacity (25% growth capacity for transformers is common practice).

• A transformer with fans increases its rating. A 1000 kVA dry-type transformer with fans is good for 1333 kVA (33% increase). For an

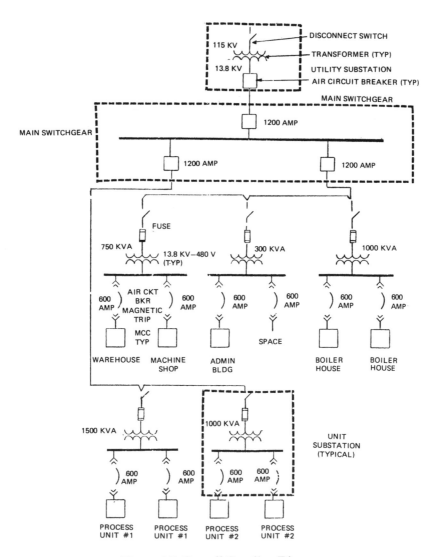

Figure 4-7. Overall One-line Diagram

oil-type transformer a factor of 25% is used. Fans should only be considered for emergency conditions or for expanding existing plants.

- The question comes up as to where to locate equipment. Incoming switchgear is usually located near the property line so that the utility company can gain easy access to the equipment. Substations and motor control centers are usually located indoors in electrical rooms.

ELECTRICAL ROOMS

The electrical engineer should keep in mind the following when specifying electrical room requirements.

- Do not allow roof penetrations. Any roof opening increases the risk of fluid entering the electrical equipment.

- Do not allow other trades to use electrical room space.

- General ambient temperature should be 40°C. (Special equipment, such as computers, may require air conditioning.)

- Layout electrical rooms with the following in mind:
  (a) Sufficient aisle space and door clearances should be provided to allow for maintenance and replacing of damaged transformers and breakers. Use the recommended clearances established by the vendor.
  (b) Double doors of adequate height (usually 8 feet) should be provided at exits in order to remove equipment.

SHORT-CIRCUIT CURRENTS

Faults occur for many reasons: deterioration of insulation, accidents, rats electrocuted across power leads, equipment failure, and a multitude of other events. When a fault occurs a large short-circuit current flows. At first it has an initial peak or asymmetrical value, but after a period of time it will become symmetrical about the zero axis. Equipment must be rated to meet both the full-load currents and the short-circuit currents available.

RATING SUBSTATION BREAKERS (480V)

All breakers should be sized to meet the full-load current and the available short-circuit current. They must also be able to coordinate with the system. Coordination of protective devices such as breakers means:

(a) That a protective device will not trip under normal operating conditions, such as when the motor is started.

(b) That the protective element closest to the fault will open before the other devices upstream.

The impedance of the transformer limits the amount of short-circuit current that could flow. Table 4-1 illustrates how the transformer rating and utility system rating affect the short-circuit current. Use this table as a guide; a more detailed analysis is required on actual selection.

SIM 4-11
For a 1000 kVA transformer, 5.75% impedance, and available primary short circuit of 250 MVA, determine the short-circuit current, assuming 100% motor contribution.

Answer
From Table 4-1, the short-circuit current is 24,400 symmetrical amperes.

Note: Usually at the beginning of a project, if no utility data are available, assume unlimited short-circuit current and 100% motor load contribution.

SIM 4-12
A 1500 kVA substation secondary breaker feeds a 600 ampere Motor Control Center bus. Select a breaker frame to meet this load.

Answer
From Table 4-1, the 1600 ampere breaker is the minimum breaker size.

Based on a continuous current rating, a 600 ampere frame breaker would have been sufficient, but the 600 ampere breaker can only handle a short-circuit current of 22,000 amperes. The 1600 ampere breaker can handle a short-circuit current up to 50,000 amperes and is good for 1600 continuous amperes. Thus, the frame size required is 1600 amperes. The trip rating that indicates when the breaker will open can be set at any

Table 4-1. Short-circuit Application Table: 480 Volts, Three-phase

Without FANS

Transformer Rating 3-Phase KVA and Impedance Percent	Maximum Short Circuit MVA Available from Primary System	Normal Load Continuous Current Amp	Short Circuit Current RMS Symmetrical Amp			Long-Time Instantaneous Recommended Min. Breaker Frame
			Transformer Alone	100% Motor Load	Combined	
1	2	3	4	5	6	7
300 5%	50	361	6500	1400	7900	225
	100		6900		8300	
	150		7000		8400	
	250		7100		8500	
	500		7200		8600	
	750		7200		8600	
	Unlimited		7300		8700	
500 5%	50	601	10000	2400	12400	225
	100		10900		13300	
	150		11300		13700	
	250		11600		14000	
	500		11800		14200	
	750		11800		14200	
	Unlimited		12000		14400	
750 5.75%	50	902	12500	3600	16100	225
	100		13900		17500	
	150		14400		18000	
	250		14900		18500	
	500		15300		18900	
	750		15400		19000	
	Unlimited		15700		19300	

|more|

1000 5.75%	50	1203	15500	4800	20300	225
	100		17800		22600	600
	150		18800		23600	
	250		19600		24400	
	500		20200		25000	
	750		20500		25300	
	Unlimited		20900		25700	
1500 5.75%	50	1804	20600	7200	27800	600
	100		24900		32100	1600
	150		26700		33900	
	250		28400		35600	
	500		29800		37000	
	750		30300		37500	
	Unlimited		31400		38600	
2000 5.75%	50	2406	24700	9600	34300	1600
	100		31100		40700	
	150		34000		43600	
	250		36700		46300	
	500		39100		48700	
	750		40000		49600	3000
	Unlimited		41900		51500	
2500 5.75%	50	3008	28000	12000	40000	1600
	100		36400		48400	
	150		40500		52500	3000
	250		44500		56500	
	500		48100		60100	
	750		49500		61500	
	Unlimited		52300		64300	
3000 5.75%	50	3607	30700	14400	45100	1600
	100		41200		55600	3000
	150		46500		60900	
	250		51900		66300	
	500		56800		71200	4000
	750		58700		73100	
	Unlimited		62700		77100	

/more/

Table 4-1. Short Circuit Application Table: 480 Volts, Three-phase (concluded)

Breaker Frame	Continuous Current Ratings Typical Trip Sizes	480 V Breaker Rating	Short-time Rating Amperes RMS Symmetrical
8	9	10	11
225	15, 20, 30, 40, 50, 70, 90, 100, 125, 150, 175, 200, 225	225	9,000
600	40, 50, 70, 90, 100, 125, 150, 175, 200, 225, 250, 300, 350, 400, 500, 600	600	22,000
1600	200, 225, 250, 275, 300, 350, 400, 500, 800, 1000, 1200, 1600	1600	50,000
3000	2000, 2500, 3000	3000	65,000
4000	2000, 2500, 3000, 4000	4000	85,000

value as indicated in Table 4-1. To protect the motor control center, a 600 ampere trip would be chosen.

ANALYZING BIDS

It is the responsibility of the electrical engineer to recommend vendors to build the electrical equipment for the plant. To get competitive bids, requests for quotations are initiated with several bidders. After the bids have been received, they must be evaluated in terms of quality, costs and schedule in particular.

- Check to see if a vendor has met specifications. Check capacities, rating, etc.

- Look at advantages and disadvantages with each vendor. Physical size, energy consumption, noise level, weight, etc. may affect the final recommendation.

- Analyze delivery schedule.

- Evaluate cost picture. Check if all items are included.

Sometimes a client may wish to direct purchase an item, due to overriding factors, such as spare parts or to match an existing installation.

JOB SIMULATION—SUMMARY PROBLEM

SIM 4-13

(a) The client wishes to know the power factor at which the plant is operating. Exclude motors below 3 HP from computations. Assume a lighting load of 40 kW. The plant is comprised of two identical modules (two motors for each equipment number listed in SIM 2-6). (Remember that kVAs at different power factors cannot be added directly.)

(b) Based on the total kVA of the plant, determine the transformer size and the rating of each breaker. Assume an individual breaker feeds each module. The substation data will be sent to the three industries listed below for competitive bids:

> Recommended vendors: ABC Industries
> DEF Industries
> GHI Industries

(c) Based on the substation quotes received in SIM 3-7(b), recommend a vendor. Required delivery date is 35 weeks from date of order.

Complete the following form:

Simplified Bid Analysis Form

	Vendor's Name	Vendor's Name	Vendor's Name
Cost per Substation			
Delivery			
Meets Specification			
Overall Area			

Bidder Recommended

Reason:

(a) Based on the motor list, the plant power factor was estimated at .87. Here's what the client expected to see:

Module #1

Lighting $kW_3 = 40$ Total

Motors 3-30	Motors 40-100
30	60
10	100
25	40
20	50
20	250
20	
5	
130	

At Full Load:

$PF = 83.5$ $PF = 87.4$

$\eta = 86.2$ $\eta = 90.9$

$$kVA_1 = \frac{130 \times .746}{.83 \times .86} = 135 \qquad kVA_2 = \frac{250 \times .746}{.90 \times .87} = 238$$

$$\begin{aligned} kW_1 &= kVA\ CoS\Theta \\ &= kVA\ .83 \\ &= 112 \end{aligned} \qquad \begin{aligned} kW_2 &= kVA_2\ CoS\Theta \\ &= kVA\ .87 \\ &= 207 \end{aligned}$$

$kVAR_1 = kVA_1\ Sin\Theta$ $kVAR_2 = kVA_2\ Sin\Theta$

$\Theta = 33°$ $\Theta = 29°$

$\begin{aligned} kVAR_1 &= kVA_1 \times .54 \\ &= 135 \times .54 \\ &= 73 \end{aligned}$ $\begin{aligned} kVAR_2 &= kVA_2 \times .48 \\ kVAR_2 &= 115 \end{aligned}$

$$\text{kW}_{\text{total}} \quad = \underset{\underset{1}{\text{Module}}}{\text{kW}_1 + \text{kW}_2} + \underset{\underset{2}{\text{Module}}}{\text{kW}_1 + \text{kW}_2} + \text{kW}_3 \quad = = 678 \text{ kW}$$

$$\text{kVAR}_{\text{total}} = \underset{\text{Module 1}}{\text{kVAR}_1 + \text{kVAR}_2} + \underset{\text{Module 2}}{\text{kVAR}_1 + \text{kVAR}_2}$$
$$= 73 + 115 + 73 + 115 = 376$$

$$\text{kVA}_{\text{total}} = \sqrt{(678)^2 + (376)^2} = 774 \text{ kVA}$$

$$\text{CoS}\Theta = \frac{\text{kW}_{\text{total}}}{\text{kVA}_{\text{total}}} = \frac{678}{774} = .87$$

(b)   The substation requisition should have included a 1000 kVA trans-
former and two 600 ampere feeder breakers.
Transformer   = 1.25 (Total kVA) = 1.25 × 774
                        = 967 kVA size required.
The closest transformer size is 1000 kVA.

The minimum breaker size, based on Table 4-1, is 600 amperes. This
size is required even though the total load on each breaker is 774 ÷ 2 = 387
kVA or

$$\frac{387 \text{ k}}{\sqrt{3} \times 480} = 466 \text{ Amps}$$

The breaker must be sized to meet the short-circuit current of approximately 25,000 amps symmetrical.

Quotes received based on the substation requisition:

ABC Industries
Dear Sir:
    In reply to subject inquiry, we are pleased to quote on the following:
    1—480 Volt Substation including:
    1.   Incoming line compartment with load interrupter switch and current
         limiting fuse
    2.   Dry type transformer 1000 kVA 13.8 kV to 480 volt wye
    3.   Low voltage switchgear with main bus, feeder instrument and meter-
         ing—2 breakers
    Equipment shall be arranged in accordance with layout attached.
    Total Price is $37,000.
    Shipment can be made 30 weeks after receipt of order.
    Standard terms 30 days from date of invoice.
                                            Very truly yours,
                                            /s/ Al B. See

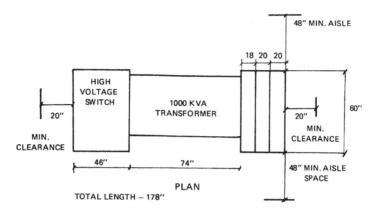

PLAN

TOTAL LENGTH — 178″

DEF Industries
Gentlemen:

We regret to inform you that at this time we cannot quote. We appreciate your interest in DEF Industries and hope that we can be of service in the future.

Very truly yours,
/s/Dee Frank

GHI Industries
Dear Sir:

We wish to offer our package completely in accordance with your specifications.

The total price for the order is $38,500.

The above prices are quoted F.O.B. factory.

The earliest time equipment can be shipped is 40 weeks after receipt of order.

The attached layout shows the arrangement of the equipment.

Yours very truly,
/s/Gee Hi Eye

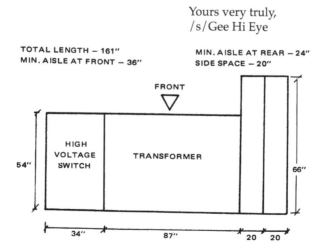

(c)   A completed bid analysis form:

	Vendor's Name	Vendor's Name	Vendor's Name
	ABC Ind.	DEF Ind.	GHI Ind.
Cost per Substation	$37,000	Declined	$38,500
Delivery	30 weeks		40 weeks
Meets Specification	Yes		Yes
Overall Area	178" × 60"		161" × 66"

Bidder
Recommended   ABC Industries

Reason: Only vendor which meets required delivery.
                Lower price.

SUMMARY

Transformer ratings are based on kVA. Since loads constantly change, a sizing based on the sum of the motor and miscellaneous loads, plus 25% extra gives a reasonable initial basis for determining capacity.

When determining the rating of equipment, always make sure that the equipment meets the short-circuit current of the plant.

The one-line diagram is the single most important drawing for the electrical design. It serves as the basis for:

Short-circuit calculations
Coordination studies
Power factor correction
Equipment selection
Power plans

In addition, the one-line diagram is a must for plant maintenance. By use of the one-line diagram, problem areas can be located and corrective action can be taken.

Chapter 5

Conduit and Conductor Sizing

CONDUCTORS FOR USE IN RACEWAYS

Once the facility environment has been established, cable insulation can be selected. Table 5-1 lists a number of insulations commonly used. Each insulation has a specified operating temperature. Types THW, RHW, and XHHW are frequently used in industrial plants for 600 volt or less distribution.

Allowable ampacities of copper conductors in conduit at 30°C are shown in Table 5-2. Sometimes the designations 0, 00, 000, and 0000 are referred to as 1/0, 2/0, 3/0, 4/0, respectively.

Tables 5-2 through 5-4 are reproduced by permission from the National Electrical Code 2008, copyright by the National Fire Protection Association. Copies of the complete code are available from the association, Batterymarch Park, Quincy, Massachusetts 02269.

SIM 5-1

For (3) 350 MCM THW copper conductors, run in conduit at 30°C, determine the conductor capacity.

Answer

From Table 5-2 (page 91) type THW is rated for 310 amperes.

SIM 5-2

What type of insulation is type RHW?

Answer

From Table 5-1 RHW is composed of moisture and heat resistant rubber.

SIM 5-3

Determine the ampacity for (3) 4/0 cross-linked polyethylene copper conductors run in conduit at 86°F in a dry location.

Table 5-1. Conductors for Use in Raceways (A Partial Listing)

TYPE	INSULATION
FEP	Fluorinated Ethylene Propylene
MI	Mineral Insulation (metal sheathed)
RH	Heat-resistant Rubber
RHH	High-heat-resistant Rubber
RHW	Moisture and Heat-resistant Rubber
RUH	Heat-resistant Latex Rubber
RUW	Moisture-resistant Latex Rubber
T	Thermoplastic
TW	Moisture-resistant Thermoplastic
THW	Moisture and Heat-resistant Thermoplastic
THWN	Moisture and Heat-resistant Thermoplastic with Nylon Jacket
THHN	High-heat -resistant Thermoplastic with Nylon Jacket
XHHW	Moisture and Heat-resistant Cross-linked Synthetic Polymer
V	Varnished Cambric
UF	Underground Feeder

NOTE: Asbestos (A) is banned from use due to carcinogenic properties.

Answer

In Table 5-2 (page 90) cross-linked polyethylene copper cable is type XHHW. When used in a dry location, it is good for 90°C. From Table 5-2, 0000 is good for 260 amperes.

CONDUCTOR DERATINGS

Derating Due to the Number of Conductors

If three conductors or less are run in a conduit, no derating need be applied. Based on the number of conductors above three, additional deratings must be used. For example: 4 to 6 conductors derate the cable 80%; 7 to 24, 70%; 25 to 42, 60%; and 43 and above, 50%. Because of these deratings and ease in pulling, many times six conductors servicing a load are run in two conduits.

Derating Due to Ambient Temperature

If the ambient differs from 30°C, another derating should be applied. These correction factors are included in Table 5-2.

SIM 5-4

What is the allowable ampacity of (6) 500 MCM, THW copper conductors run in the same conduit at 50°C ambient?

Answer

From Table 5-2 (page 91), the allowable ampacity of 500 MCM cable is 380 amperes.

Derating factor for six conductors is .8; derating for 50°C is .75.

Allowable ampacity =. (380 × .8 × .75) = 228 amps

Observation

It should be noted from Table 5-2 that some conductor sizes are not practical. Doubling the area of a conductor, i.e., 500 MCM cable to a 1000 MCM cable, does not double the ampacity rating. In fact, doubling the area only results in 50% more capacity. Using conductors larger than 500 MCM is also not recommended, due to difficulties in installation. Consequently, if more capacity is required than can be provided by a single 500 MCM conductor, multiple conductors are used in parallel.

CABLE SIZING

Sizing a Feeder to a Motor for Continuous Service

The minimum size cable for power conductors is #12. The cable capacity for a motor should be equal to 1.25 × the full load amperes of the motor. Typical full-load currents are illustrated in Chapter 3, Table 3-2d.

SIM 5-5

What cable size should be used for a 50 HP induction motor? Assume the cable is type THW and run in conduit. Ambient temperature is 30°C.

Answer

From Table 3-2d (page 48), Chapter 3, a 50 HP motor has an FLA (full load amperes) of 65 amps. Cable must be equal to or above 1.25 × 65 = 81 amps. Use #4 AWG (from Table 3-2, page 90).

Sizing Feeder to Several Motors

The size of the feeder which has more than one motor unit is based on 1.25 times the full-load current FLA of the largest motor, plus the full-load current of the others. Note: If two or more motors "tie" for the largest, multiply only the FLA of one by 1.25.

Reprinted with permission from NFPA 70-2008, National Electrical Code®, Copyright© 2007, National Fire Protection Association, Quincy, Massachusetts 02269. This reprinted material is not the complete and official position of the N2FPA on the referenced subject, which is represented only by the standard in its entirety. National Electrical Code® and NEC® are Registered Trademarks of the National Fire Protection Association, Inc., Quincy, MA.

Table 5-2.

Table 310.16 Allowable Ampacities of Insulated Conductors Rated 0 Through 2000 Volts, 60°C Through 90°C (140°F Through 194°F), Not More Than Three Current-Carrying Conductors in Raceway, Cable, or Earth (Directly Buried), Based on Ambient Temperature of 30°C (86°F)

| | Temperature Rating of Conductor [See Table 310.13(XX).] | | | | | | |
| | COPPER | | | ALUMINUM OR COPPER-CLAD ALUMINUM | | | |
Size AWG or kcmil	60°C (140°F) Types TW, UF	75°C (167°F) Types RHW, THW, THHW, THWN, XHHW, USE, ZW	90°C (194°F) Types TBS, SA, SIS, FEP, FEPB, MI, RHH, RHW-2, THHN, THHW, THW-2, THWN-2, USE-2, XHH, XHHW, XHHW-2, ZW-2	60°C (140°F) Types TW, UF	75°C (167°F) Types RHW, THW, THWN, XHHW, USE	90°C (194°F) Types TBS, SA, SIS, THHN, THHW, THW-2, THWN-2, RHH, RHW-2, USE-2, XHH, XHHW, XHHW-2, ZW-2	Size AWG or kcmil
18	—	—	14	—	—	—	—
16	—	—	18	—	—	—	—
14*	20	20	25	—	—	—	—
12*	25	25	30	20	20	25	12*
10*	30	35	40	25	30	35	10*
8	40	50	55	30	40	45	8
6	55	65	75	40	50	60	6
4	70	85	95	55	65	75	4
3	85	100	110	65	75	85	3
2	95	115	130	75	90	100	2
1	110	130	150	85	100	115	1
1/0	125	150	170	100	120	135	1/0
2/0	145	175	195	115	135	150	2/0
3/0	165	200	225	130	155	175	3/0
4/0	195	230	260	150	180	205	4/0

Table 5-2. (Continued)

Size							Size
250	215	255	290	170	205	230	250
300	240	285	320	190	230	255	300
350	260	310	350	210	250	280	350
400	280	335	380	225	270	305	400
500	320	380	430	260	310	350	500
600	355	420	475	285	340	385	600
700	385	460	520	310	375	420	700
750	400	475	535	320	385	435	750
800	410	490	555	330	395	450	800
900	435	520	585	355	425	480	900
1000	455	545	615	375	445	500	1000
1250	495	590	665	405	485	545	1250
1500	520	625	705	435	520	585	1500
1750	545	650	735	455	545	615	1750
2000	560	665	750	470	560	630	2000

CORRECTION FACTORS

For ambient temperatures other than 30°C (86°F), multiply the allowable ampacities shown above by the appropriate factor shown below.

Ambient Temp. (°C)							Ambient Temp. (°F)
21–25	1.08	1.05	1.08	1.04	1.05	1.04	70–77
26–30	1.00	1.00	1.00	1.00	1.00	1.00	78–86
31–35	0.91	0.94	0.91	0.96	0.94	0.96	87–95
36–40	0.82	0.88	0.82	0.91	0.88	0.91	96–104
41–45	0.71	0.82	0.71	0.87	0.82	0.87	105–113
46–50	0.58	0.75	0.58	0.82	0.75	0.82	114–122
51–55	0.41	0.67	0.41	0.76	0.67	0.76	123–131
56–60	—	0.58	—	0.71	0.58	0.71	132–140
61–70	—	0.33	—	0.58	0.33	0.58	141–158
71–80	—	—	—	0.41	—	0.41	159–176

* See 240.4(D).

Table 5-2. (Continued)

Table 310.17 Allowable Ampacities of Single-Insulated Conductors Rated 0 Through 2000 Volts in Free Air, Based on Ambient Air Temperature of 30°C (86°F)

Size AWG or kcmil	Temperature Rating of Conductor [See Table 310.13.]						Size AWG or kcmil
	COPPER			ALUMINUM OR COPPER-CLAD ALUMINUM			
	60°C (140°F)	75°C (167°F)	90°C (194°F)	60°C (140°F)	75°C (167°F)	90°C (194°F)	
	Types TW, UF	Types RHW, THW, THHW, THWN, XHHW, ZW	Types TBS, SA, SIS, FEP, FEPB, MI, RHH, RHW-2, THHN, THHW, THW-2, THWN-2, USE-2, XHH, XHHW, XHHW-2, ZW-2	Types TW, UF	Types RHW, THHW, THW, THWN, XHHW	Types TBS, SA, SIS, THHN, THHW, THW-2, THWN-2, RHH, RHW-2, USE-2, XHH, XHHW, XHHW-2, ZW-2	
18	—	—	18				
16	—	—	24				
14*	25	30	35				
12*	30	35	40	25	30	35	12*
10*	40	50	55	35	40	40	10*
8	60	70	80	45	55	60	8
6	80	95	105	60	75	80	6
4	105	125	140	80	100	110	4
3	120	145	165	95	115	130	3
2	140	170	190	110	135	150	2
1	165	195	220	130	155	175	1
1/0	195	230	260	150	180	205	1/0
2/0	225	265	300	175	210	235	2/0
3/0	260	310	350	200	240	275	3/0
4/0	300	360	405	235	280	315	4/0
250	340	405	455	265	315	355	250
300	375	445	505	290	350	395	300
350	420	505	570	330	395	445	350

Table 5-2. (Continued)

400 500	455 515	545 620	615 700	355 405	425 485	480 545	400 500
600	575	690	780	455	540	615	600
700	630	755	855	500	595	675	700
750	655	785	885	515	620	700	750
800	680	815	920	535	645	725	800
900	730	870	985	580	700	785	900
1000	780	935	1055	625	750	845	1000
1250	890	1065	1200	710	855	960	1250
1500	980	1175	1325	795	950	1075	1500
1750	1070	1280	1445	875	1050	1185	1750
2000	1155	1385	1560	960	1150	1335	2000

CORRECTION FACTORS

For ambient temperatures other than 30°C (86°F), multiply the allowable ampacities shown above by the appropriate factor shown below.

Ambient Temp. (°C)							Ambient Temp. (°F)
21–25	1.08	1.05	1.04	1.08	1.05	1.04	70–77
26–30	1.00	1.00	1.00	1.00	1.00	1.00	78–86
31–35	0.91	0.94	0.96	0.91	0.94	0.96	87–95
36–40	0.82	0.88	0.91	0.82	0.88	0.91	96–104
41–45	0.71	0.82	0.87	0.71	0.82	0.87	105–113
46–50	0.58	0.75	0.82	0.58	0.75	0.82	114–122
51–55	0.41	0.67	0.76	0.41	0.67	0.76	123–131
56–60	—	0.58	0.71	—	0.58	0.71	132–140
61–70	—	0.33	0.58	—	0.33	0.58	141–158
71–80	—	—	0.41	—	—	0.41	159–176

* See 240.4(D).

Table 5-2. (Continued)

Table 310.18 Allowable Ampacities of Insulated Conductors Rated 0 Through 2000 Volts, 150°C Through 250°C (302°F Through 482°F). Not More Than Three Current-Carrying Conductors in Raceway or Cable, Based on Ambient Air Temperature of 40°C (104°F)

Size AWG or kcmil	Temperature Rating of Conductor [See Table 310.13(A).]				Size AWG or kcmil
	150°C (302°F)	200°C (392°F)	250°C (482°F)	150°C (302°F)	
	Type Z	Types FEP, FEPB, PFA, SA	Types PFAH, TFE	Type Z	
	COPPER		NICKEL OR NICKEL-COATED COPPER	ALUMINUM OR COPPER-CLAD ALUMINUM	
14	34	36	39	—	14
12	43	45	54	30	12
10	55	60	73	44	10
8	76	83	93	57	8
6	96	110	117	75	6
4	120	125	148	94	4
3	143	152	166	109	3
2	160	171	191	124	2
1	186	197	215	145	1
1/0	215	229	244	169	1/0
2/0	251	260	273	198	2/0
3/0	288	297	308	227	3/0
4/0	332	346	361	260	4/0

Table 5-2. (Continued)

CORRECTION FACTORS

Ambient Temp. (°C)	For ambient temperatures other than 40°C (104°F), multiply the allowable ampacities shown above by the appropriate factor shown below.			Ambient Temp. (°F)
41–50	0.95	0.97	0.98	0.95	105–122
51–60	0.90	0.94	0.95	0.90	123–140
61–70	0.85	0.90	0.93	0.85	141–158
71–80	0.80	0.87	0.90	0.80	159–176
81–90	0.74	0.83	0.87	0.74	177–194
91–100	0.67	0.79	0.85	0.67	195–212
101–120	0.52	0.71	0.79	0.52	213–248
121–140	0.30	0.61	0.72	0.30	249–284
141–160	—	0.50	0.65	—	285–320
161–180	—	0.35	0.58	—	321–356
181–200	—	—	0.49	—	357–392
201–225	—	—	0.35	—	393–437

Table 5-2. (Continued)

Table 310.19 Allowable Ampacities of Single-Insulated Conductors, Rated 0 Through 2000 Volts, 150°C Through 250°C (302°F Through 482°F), in Free Air, Based on Ambient Air Temperature of 40°C (104°F)

	Temperature Rating of Conductor [See Table 310.13(A).]				
	150°C (302°F)	200°C (392°F)	250°C (482°F)	150°C (302°F)	
	Type Z	Types FEP, FEPB, PFA, SA	Types PFAH, TFE	Type Z	
Size AWG or kcmil	COPPER	COPPER	NICKEL, OR NICKEL-COATED COPPER	ALUMINUM OR COPPER-CLAD ALUMINUM	Size AWG or kcmil
14	46	54	59	—	14
12	60	68	78	47	12
10	80	90	107	63	10
8	106	124	142	83	8
6	155	165	205	112	6
4	190	220	278	148	4
3	214	252	327	170	3
2	255	293	381	198	2
1	293	344	440	228	1
1/0	339	399	532	263	1/0
2/0	390	467	591	305	2/0
3/0	451	546	708	351	3/0
4/0	529	629	830	411	4/0

Table 5-2. (Continued)

CORRECTION FACTORS

Ambient Temp. (°C)	For ambient temperatures other than 40°C (104°F), multiply the allowable ampacities shown above by the appropriate factor shown below.			Ambient Temp. (°F)
41–50	0.95	0.97	0.98	0.95	105–122
51–60	0.90	0.94	0.95	0.90	123–140
61–70	0.85	0.90	0.93	0.85	141–158
71–80	0.80	0.87	0.90	0.80	159–176
81–90	0.74	0.83	0.87	0.74	177–194
91–100	0.67	0.79	0.85	0.67	195–212
101–120	0.52	0.71	0.79	0.52	213–248
121–140	0.30	0.61	0.72	0.30	249–284
141–160	—	0.50	0.65	—	285–320
161–180	—	0.35	0.58	—	321–356
181–200	—	—	0.49	—	357–392
201–225	—	—	0.35	—	393–437

Table 5-2. (Continued)

Table 310.20 Ampacities of Not More Than Three Single Insulated Conductors, Rated 0 Through 2000 Volts, Supported on a Messenger, Based on Ambient Air Temperature of 40°C (104°F)

Size AWG or kcmil	Temperature Rating of Conductor [See Table 310.13(A).]				Size AWG or kcmil
	COPPER		ALUMINUM OR COPPER-CLAD ALUMINUM		
	75°C (167°F)	90°C (194°F)	75°C (167°F)	90°C (194°F)	
	Types RHW, THHW, THW, THWN, XHHW, ZW	Types MI, THHN, THHW, THW-2, THWN-2, RHH, RHW-2, USE-2, XHHW, XHHW-2, ZW-2	Types RHW, THW, THWN, THHW, XHHW	Types THHN, THHW, RHH, XHHW, RHW-2, XHHW-2, THW-2, THWN-2, USE-2, ZW-2	
8	57	66	44	51	8
6	76	89	59	69	6
4	101	117	78	91	4
3	118	138	92	107	3
2	135	158	106	123	2
1	158	185	123	144	1
1/0	183	214	143	167	1/0
2/0	212	247	165	193	2/0
3/0	245	287	192	224	3/0
4/0	287	335	224	262	4/0
250	320	374	251	292	250
300	359	419	282	328	300
350	397	464	312	364	350
400	430	503	339	395	400
500	496	580	392	458	500
600	553	647	440	514	600
700	610	714	488	570	700
750	638	747	512	598	750
800	660	773	532	622	800
900	704	826	572	669	900
1000	748	879	612	716	1000

Table 5-2. (Continued)

CORRECTION FACTORS

Ambient Temp. (°C)	For ambient temperatures other than 40°C (104°F), multiply the allowable ampacities shown above by the appropriate factor shown below.			Ambient Temp. (°F)
21–25	1.20	1.20	1.14	70–77
26–30	1.13	1.13	1.10	79–86
31–35	1.07	1.07	1.05	88–95
36–40	1.00	1.00	1.00	97–104
41–45	0.93	0.93	0.95	106–113
46–50	0.85	0.85	0.89	115–122
51–55	0.76	0.76	0.84	124–131
56–60	0.65	0.65	0.77	133–140
61–70	0.38	0.38	0.63	142–158
71–80	—	—	0.45	160–176

Table 5-2. (Concluded)

Table 310.21 Ampacities of Bare or Covered Conductors in Free Air, Based on 40°C (104°F) Ambient, 80°C (176°F) Total Conductor Temperature, 610 mm/sec (2 ft/sec) Wind Velocity

Copper Conductors				AAC Aluminum Conductors			
Bare		Covered		Bare		Covered	
AWG or kcmil	Amperes	AWG or kcmil	Amperes	AWG or kcmil	Amperes	AWG or kcmil	Amperes
8	98	8	103	8	76	8	80
6	124	6	130	6	96	6	101
4	155	4	163	4	121	4	127
2	209	2	219	2	163	2	171
1/0	282	1/0	297	1/0	220	1/0	231
2/0	329	2/0	344	2/0	255	2/0	268
3/0	382	3/0	401	3/0	297	3/0	312
4/0	444	4/0	466	4/0	346	4/0	364
250	494	250	519	266.8	403	266.8	423
300	556	300	584	336.4	468	336.4	492
500	773	500	812	397.5	522	397.5	548
750	1000	750	1050	477.0	588	477.0	617
1000	1193	1000	1253	556.5	650	556.5	682
—	—	—	—	636.0	709	636.0	744
—	—	—	—	795.0	819	795.0	860
—	—	—	—	954.0	920	—	—
—	—	—	—	1033.5	968	1033.5	1017
—	—	—	—	1272	1103	1272	1201
—	—	—	—	1590	1267	1590	1381
—	—	—	—	2000	1454	2000	1527

SIM 5-6

A power feeder supplies a 25-HP motor and two 30-HP motors. Determine the size of the feeder if it is rated for ambient 30°C and cable run in conduit. (Power Feeder, Cu, THW.)

Answer

FLA – 30 HP motor – 40 amps (from Table 3-2d).

FLA – 25 HP motor – 34 amps (from Table 3-2d).

Cable must meet: $(1.25 \times 40) + 34 + 40 = 124$ amps.

The nearest cable size from Table 5-2 (page 90) is #1 THW.

SIZING A CONDUIT

Once the conductors are selected, they must be protected from physical damage through the use of an enclosed conduit. Depending on the application the material used for the conduit can be rigid steel, electrical metallic tubing (EMT), flexible metal or PVC. Note that for metallic materials, the conduit often serves as the equipment ground.

The size of a conduit depends on the allowable percent fill of the conduit area. Table 5-3 illustrates allowable fills. Table 5-4 is based on Table 5-3 and enables the engineer to readily select the number of conductors which can be installed in a conduit.

Table 5-3.
Table 1 Percent of Cross Section of Conduit and Tubing for Conductors

Number of Conductors	All Conductor Types
1	53
2	31
Over 2	40

FPN No. 1: Table 1 is based on common conditions of proper cabling and alignment of conductors where the length of the pull and the number of bends are within reasonable limits. It should be recognized that, for certain conditions, a larger size conduit or a lesser conduit fill should be considered.

FPN No. 2: When pulling three conductors or cables into a raceway, if the ratio of the raceway (inside diameter) to the conductor or cable (outside diameter) is between 2.8 and 3.2, jamming can occur. While jamming can occur when pulling four or more conductors or cables into a raceway, the probability is very low.

Annex C, from the 2008 National Electric Code®, is duplicated at the end of this chapter. These tables stipulate how many conductors can be run in a conduit based on the allowable fills mentioned in Table 5-3.

SIM 5-7

What is the allowable percent fill for three conductors (type THW) run in conduit?

Answer

From Table 5-3, 40%.

SIM 5-8

A control cable consisting of thirty-seven #14 wires (type TW) is run from MCC No. 1 to Process Panel No. 2. What size conduit is required?

Answer

From Table 5-3 the maximum conduit fill is 40%. From the Annex C table on page 108, the answer can be read directly. Since 40 wires can be run in a 1-1/4″ conduit, the correct size is 1-1/4″.

Power and Control Run in the Same Conduit

Frequently the electrical engineer is faced with the problem of sizing a conduit for both power and control wires. An example might be a local STOP-START pushbutton at the motor. The control wires would be run in the same conduit as the power leads if the horsepower is 60 HP or less. Above this size, it becomes impractical to pull the smaller control wires with the larger power conductors. For different size cables, first use the Annex C tables on pages 000-000 to determine the area of each cable. Then look at Table 5-4 under the appropriate percent fill column and choose a conduit size whose area is equal to or greater than the total conductor area. Note that control wire insulation must be rated for high voltage if high voltage power wiring is present.

SIM 5-9

Determine the conduit size for the 50 HP motor, in SIM 5-5. Assume a local stop-start pushbutton requiring three #14 control cables.

Answer

From SIM 5-5, #4 conductors are required for a 50 HP motor. Cable manufacturers' tables show cable areas for single conductors. Multiply the area by 3 to get the total area. This approximation gives a conservative

Table 5-4.

Reprinted with permission from NFPA 70-2008, National Electrical Code®, Copyright©2007, National Fire Protection Association, Quincy, Massachusetts 02269. This reprinted material is not the complete and official position of the NFPA on the referenced subject, which is represented only by the standard in its entirety.

Table 4 Dimensions and Percent Area of Conduit and Tubing
(Areas of Conduit or Tubing for the Combinations of Wires Permitted in Table 1, Chapter 9)

Article 358 — Electrical Metallic Tubing (EMT)

Metric Designator	Trade Size	Nominal Internal Diameter mm	in.	Total Area 100% mm²	in.²	60% mm²	in.²	1 Wire 53% mm²	in.²	2 Wires 31% mm²	in.²	Over 2 Wires 40% mm²	in.²
16	½	15.8	0.622	196	0.304	118	0.182	104	0.161	61	0.094	78	0.122
21	¾	20.9	0.824	343	0.533	206	0.320	182	0.283	106	0.165	137	0.213
27	1	26.6	1.049	556	0.864	333	0.519	295	0.458	172	0.268	222	0.346
35	1¼	35.1	1.380	968	1.496	581	0.897	513	0.793	300	0.464	387	0.598
41	1½	40.9	1.610	1314	2.036	788	1.221	696	1.079	407	0.631	526	0.814
53	2	52.5	2.067	2165	3.356	1299	2.013	1147	1.778	671	1.040	866	1.342
63	2½	69.4	2.731	3783	5.858	2270	3.515	2005	3.105	1173	1.816	1513	2.343
78	3	85.2	3.356	5701	8.846	3421	5.307	3022	4.688	1767	2.742	2280	3.538
91	3½	97.4	3.834	7451	11.545	4471	6.927	3949	6.119	2310	3.579	2980	4.618
103	4	110.1	4.334	9521	14.753	5712	8.852	5046	7.819	2951	4.573	3808	5.901

Article 362 — Electrical Nonmetallic Tubing (ENT)

Metric Designator	Trade Size	Nominal Internal Diameter mm	in.	Total Area 100% mm²	in.²	60% mm²	in.²	1 Wire 53% mm²	in.²	2 Wires 31% mm²	in.²	Over 2 Wires 40% mm²	in.²
16	½	14.2	0.560	158	0.246	95	0.148	84	0.131	49	0.076	63	0.099
21	¾	19.3	0.760	293	0.454	176	0.272	155	0.240	91	0.141	117	0.181
27	1	25.4	1.000	507	0.785	304	0.471	269	0.416	157	0.243	203	0.314
35	1¼	34.0	1.340	908	1.410	545	0.846	481	0.747	281	0.437	363	0.564
41	1½	39.9	1.570	1250	1.936	750	1.162	663	1.026	388	0.600	500	0.774
53	2	51.3	2.020	2067	3.205	1240	1.923	1095	1.699	641	0.993	827	1.282
63	2½	—	—	—	—	—	—	—	—	—	—	—	—
78	3	—	—	—	—	—	—	—	—	—	—	—	—
91	3½	—	—	—	—	—	—	—	—	—	—	—	—

Article 348 — Flexible Metal Conduit (FMC)

Metric Designator	Trade Size	Nominal Internal Diameter mm	in.	Total Area 100% mm²	in.²	60% mm²	in.²	1 Wire 53% mm²	in.²	2 Wires 31% mm²	in.²	Over 2 Wires 40% mm²	in.²
12	⅜	9.7	0.384	74	0.116	44	0.069	39	0.061	23	0.036	30	0.046
16	½	16.1	0.635	204	0.317	122	0.190	108	0.168	63	0.098	81	0.127
21	¾	20.9	0.824	343	0.533	206	0.320	182	0.283	106	0.165	137	0.213
27	1	25.9	1.020	527	0.817	316	0.490	279	0.433	163	0.253	211	0.327
35	1¼	32.4	1.275	824	1.277	495	0.766	437	0.677	256	0.396	330	0.511
41	1½	39.1	1.538	1201	1.858	720	1.115	636	0.985	372	0.576	480	0.743
53	2	51.8	2.040	2107	3.269	1264	1.961	1117	1.732	653	1.013	843	1.307
63	2½	63.5	2.500	3167	4.909	1900	2.945	1678	2.602	982	1.522	1267	1.963
78	3	76.2	3.000	4560	7.069	2736	4.241	2417	3.746	1414	2.191	1824	2.827
91	3½	88.9	3.500	6207	9.621	3724	5.773	3290	5.099	1924	2.983	2483	3.848
103	4	101.6	4.000	8107	12.566	4864	7.540	4297	6.660	2513	3.896	3243	5.027

cable sizing.

#4 AWA Power	$3 \times .1087 = .3261$
#14 AWA Control	$3 \times .0206 = \underline{.0618}$
	Total area .3879

From Table 5-4, based on 40% fill, conduit size is 1-1/4".

SUMMARY OF DATA

At the beginning of each project a table should be established summarizing all conduit, cable, fuse and switch or breaker sizes for each motor horsepower. Each motor and its associated auxiliaries can be determined at a glance. On a large project this type of table saves a considerable amount of time.

POWER LAYOUTS

Power layouts are drawn to scale (usually 1/4" = 1'). Conduits are grouped together where possible to form conduit banks. Usually conduits run vertically and horizontally. If conduit and cable sizes cannot be shown on drawing, a separate conduit and cable schedule is required. Conduit layouts should be coordinated with other groups (Piping, HVAC) to avoid interferences.

SIM 5-10

For the plan below, draw a conduit layout.

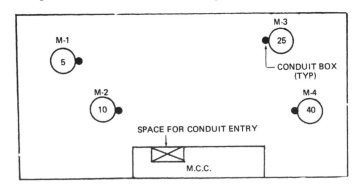

Answer

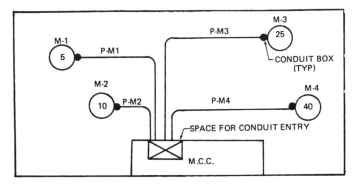

JOB SIMULATION - SUMMARY PROBLEM

SIM 5-11

(a) With the motor data of 81M 2-6 (Chapter 2) establish a conduit and cable schedule (ambient 40°C and wire type THW. Assume each stop-start motor requires three # 14 for a local pushbutton station. No local control is provided for the two-speed motor, the reversing motors, and AG-1 and FP4. Assume hoist control by others.

When power and control are run in the same conduit, designate the conduit by PC-motor number. For power alone, use P-motor number. For control alone, use C-motor number. Exclude single-phase motors from your list. Use #14 for control.

Remember control for motors above 60 HP will be run separately. Note: Cable size 3 is not frequently used.

(b) Determine the cable and conduit size required for the feed from the substation to the MCC Assume 40°C ambient.

Analysis

(a) See the conduit schedule on the following page.

(b) The cable to the MCC should be at least 1.25% of the FLA of the largest motor, plus the full load amps of the others. In this case it is approximately 525 amps. The lighting and fractional HP load of approximately IS amps is then added, bringing the total to 540 amps.

Two 350 MCM run in separate conduits can feed a total load of 2 × 310 × .88 = 545.

For this application, two 2-1/2 conduits are required.

(a)   Conduit Schedule

Conduit No.	No.	Size	Conduit Size
P-AG1	1	#2	1-1/4"
P-CF3	1	#2/0	1-1/2"
P-FP4	1	#6	1"
PC-TP-5	1	#12 ⎫	3/4"
	1	#14 ⎭	
PC-CTP6	1	#6 ⎫	1"
	1	#14 ⎭	
P-CT9	2	#8	3/4"
PC-HF10	1	#4 ⎫	3/4"
	1	#14 ⎭	
PC-HF11	1	#8 ⎫	1"
	1	#14 ⎭	
PC-BC13	1	#2 ⎫	1-1/4"
	1	#14 ⎭	
P-C16	1	#8	3/4"
P-H17	1	#12	1/2"
PC-SC19	1	#12 ⎫	3/4"
	1	#14 ⎭	

SUMMARY

The National Electrical Code gives specific tables for calculating ampacities and conduit sizes for different cables and various conditions.

In this chapter, conduit and cable determinations have been illustrated, but there are other raceway applications. Wireways and trays are used as a raceway to carry the cable. Appropriate derating factors should be applied, based on the National Electrical Code, when using other types of raceways.

The following tables are reprinted with permission from the NFPA 70-2008 National Electricl Code®. Copyright©2007, National Fire Protection Association, Quincy, Massachusetts 02269. This reprinted material is not the complete and official position of the NFPA on the referenced subject, which is represented only by the standard in its entirety. **Page numbers for the index to Annex C have been changed to reflect the numbering in this book.**

Annex C   Conduit and Tubing Fill Tables for Conductors and Fixture Wires of the Same Size

This annex is not a part of the requirements of this NFPA document but is included for informational purposes only.

*Where this table is used in conjunction with Tables C.1 through C.12, the conductors installed must be of the compact type.

Table C.1 Maximum Number of Conductors or Fixture Wires in Electrical Metallic Tubing (EMT) (*Based on Table 1, Chapter 9*)

		CONDUCTORS								
	Conductor Size	Metric Designator (Trade Size)								
Type	(AWG kcmil)	16 (½)	21 (¾)	27 (1)	35 (1¼)	41 (1½)	53 (2)	63 (2½)	78 (3)	91 (3½)	103 (4)
RHH,	14	4	7	11	20	27	46	80	120	157	201
RHW,	12	3	6	9	17	23	38	66	100	131	167
RHW-2	10	2	5	8	13	18	30	53	81	105	135
	8	1	2	4	7	9	16	28	42	55	70
	6	1	1	3	5	8	13	22	34	44	56
	4	1	1	2	4	6	10	17	26	34	44
	3	1	1	1	4	5	9	15	23	30	38
	2	1	1	1	3	4	7	13	20	26	33
	1	0	1	1	1	3	5	9	13	17	22
	1/0	0	1	1	1	2	4	7	11	15	19
	2/0	0	1	1	1	2	4	6	10	13	17
	3/0	0	0	1	1	1	3	5	8	11	14
	4/0	0	0	1	1	1	3	5	7	9	12
	250	0	0	0	1	1	1	3	5	7	9
	300	0	0	0	1	1	1	3	5	6	8
	350	0	0	0	1	1	1	3	4	6	7
	400	0	0	0	1	1	1	2	4	5	7
	500	0	0	0	0	1	1	2	3	4	6
	600	0	0	0	0	1	1	1	3	4	5
	700	0	0	0	0	0	1	1	2	3	4
	750	0	0	0	0	0	1	1	2	3	4
	800	0	0	0	0	0	1	1	2	3	4
	900	0	0	0	0	0	1	1	1	3	3
	1000	0	0	0	0	0	1	1	1	2	3
	1250	0	0	0	0	0	0	1	1	1	2
	1500	0	0	0	0	0	0	1	1	1	1
	1750	0	0	0	0	0	0	1	1	1	1
	2000	0	0	0	0	0	0	1	1	1	1
TW	14	8	15	25	43	58	96	168	254	332	424
	12	6	11	19	33	45	74	129	195	255	326
	10	5	8	14	24	33	55	96	145	190	243
	8	2	5	8	13	18	30	53	81	105	135
RHH*, RHW*, RHW-2*, THHW, THW, THW-2	14	6	10	16	28	39	64	112	169	221	282
RHH*, RHW*, RHW-2*, THHW, THW	12	4	8	13	23	31	51	90	136	177	227
	10	3	6	10	18	24	40	70	106	138	177
RHH*, RHW*, RHW-2*, THHW, THW, THW-2	8	1	4	6	10	14	24	42	63	83	106

Table C.1 *Continued*

		CONDUCTORS									
	Conductor Size	Metric Designator (Trade Size)									
Type	(AWG kcmil)	16 (½)	21 (¾)	27 (1)	35 (1¼)	41 (1½)	53 (2)	63 (2½)	78 (3)	91 (3½)	103 (4)
RHH*, RHW*, RHW-2*, TW, THW, THHW, THW-2	6	1	3	4	8	11	18	32	48	63	81
	4	1	1	3	6	8	13	24	36	47	60
	3	1	1	3	5	7	12	20	31	40	52
	2	1	1	2	4	6	10	17	26	34	44
	1	1	1	1	3	4	7	12	18	24	31
	1/0	0	1	1	2	3	6	10	16	20	26
	2/0	0	1	1	1	3	5	9	13	17	22
	3/0	0	1	1	1	2	4	7	11	15	19
	4/0	0	0	1	1	1	3	6	9	12	16
	250	0	0	1	1	1	3	5	7	10	13
	300	0	0	1	1	1	2	4	6	8	11
	350	0	0	0	1	1	1	4	6	7	10
	400	0	0	0	1	1	1	3	5	7	9
	500	0	0	0	1	1	1	3	4	6	7
	600	0	0	0	1	1	1	2	3	4	6
	700	0	0	0	0	1	1	1	3	4	5
	750	0	0	0	0	1	1	1	3	4	5
	800	0	0	0	0	1	1	1	3	3	5
	900	0	0	0	0	0	1	1	2	3	4
	1000	0	0	0	0	0	1	1	2	3	4
	1250	0	0	0	0	0	1	1	1	2	3
	1500	0	0	0	0	0	1	1	1	1	2
	1750	0	0	0	0	0	0	1	1	1	2
	2000	0	0	0	0	0	0	1	1	1	1
THHN, THWN, THWN-2	14	12	22	35	61	84	138	241	364	476	608
	12	9	16	26	45	61	101	176	266	347	443
	10	5	10	16	28	38	63	111	167	219	279
	8	3	6	9	16	22	36	64	96	126	161
	6	2	4	7	12	16	26	46	69	91	116
	4	1	2	4	7	10	16	28	43	56	71
	3	1	1	3	6	8	13	24	36	47	60
	2	1	1	3	5	7	11	20	30	40	51
	1	1	1	1	4	5	8	15	22	29	37
	1/0	1	1	1	3	4	7	12	19	25	32
	2/0	0	1	1	2	3	6	10	16	20	26
	3/0	0	1	1	1	3	5	8	13	17	22
	4/0	0	1	1	1	2	4	7	11	14	18
	250	0	0	1	1	1	3	6	9	11	15
	300	0	0	1	1	1	3	5	7	10	13
	350	0	0	1	1	1	2	4	6	9	11
	400	0	0	0	1	1	1	4	6	8	10
	500	0	0	0	1	1	1	3	5	6	8
	600	0	0	0	1	1	1	2	4	5	7
	700	0	0	0	1	1	1	2	3	4	6
	750	0	0	0	0	1	1	1	3	4	5
	800	0	0	0	0	1	1	1	3	4	5
	900	0	0	0	0	1	1	1	3	3	4
	1000	0	0	0	0	1	1	1	2	3	4
FEP, FEPB, PFA, PFAH, TFE	14	12	21	34	60	81	134	234	354	462	590
	12	9	15	25	43	59	98	171	258	337	430
	10	6	11	18	31	42	70	122	185	241	309
	8	3	6	10	18	24	40	70	106	138	177
	6	2	4	7	12	17	28	50	75	98	126
	4	1	3	5	9	12	20	35	53	69	88
	3	1	2	4	7	10	16	29	44	57	73
	2	1	1	3	6	8	13	24	36	47	60

(Continues)

Table C.1 *Continued*

		CONDUCTORS									
	Conductor Size	Metric Designator (Trade Size)									
Type	(AWG kcmil)	16 (½)	21 (¾)	27 (1)	35 (1¼)	41 (1½)	53 (2)	63 (2½)	78 (3)	91 (3½)	103 (4)
PFA, PFAH, TFE	1	1	1	2	4	6	9	16	25	33	42
PFAH, TFE PFA, PFAH, TFE, Z	1/0	1	1	1	3	5	8	14	21	27	35
	2/0	0	1	1	3	4	6	11	17	22	29
	3/0	0	1	1	2	3	5	9	14	18	24
	4/0	0	1	1	1	2	4	8	11	15	19
Z	14	14	25	41	72	98	161	282	426	556	711
	12	10	18	29	51	69	114	200	302	394	504
	10	6	11	18	31	42	70	122	185	241	309
	8	4	7	11	20	27	44	77	117	153	195
	6	3	5	8	14	19	31	54	82	107	137
	4	1	3	5	9	13	21	37	56	74	94
	3	1	2	4	7	9	15	27	41	54	69
	2	1	1	3	6	8	13	22	34	45	57
	1	1	1	2	4	6	10	18	28	36	46
XHH, XHHW, XHHW-2, ZW	14	8	15	25	43	58	96	168	254	332	424
	12	6	11	19	33	45	74	129	195	255	326
	10	5	8	14	24	33	55	96	145	190	243
	8	2	5	8	13	18	30	53	81	105	135
	6	1	3	6	10	14	22	39	60	78	100
	4	1	2	4	7	10	16	28	43	56	72
	3	1	1	3	6	8	14	24	36	48	61
	2	1	1	3	5	7	11	20	31	40	51
XHH, XHHW, XHHW-2	1	1	1	1	4	5	8	15	23	30	38
	1/0	1	1	1	3	4	7	13	19	25	32
	2/0	0	1	1	2	3	6	10	16	21	27
	3/0	0	1	1	1	3	5	9	13	17	22
	4/0	0	1	1	1	2	4	7	11	14	18
	250	0	0	1	1	1	3	6	9	12	15
	300	0	0	1	1	1	3	5	8	10	13
	350	0	0	1	1	1	2	4	7	9	11
	400	0	0	0	1	1	1	4	6	8	10
	500	0	0	0	1	1	1	3	5	6	8
	600	0	0	0	1	1	1	2	4	5	6
	700	0	0	0	0	1	1	2	3	4	6
	750	0	0	0	0	1	1	1	3	4	5
	800	0	0	0	0	1	1	1	3	4	5
	900	0	0	0	0	1	1	1	3	3	4
	1000	0	0	0	0	0	1	1	2	3	4
	1250	0	0	0	0	0	1	1	1	2	3
	1500	0	0	0	0	0	1	1	1	1	3
	1750	0	0	0	0	0	0	1	1	1	2
	2000	0	0	0	0	0	0	1	1	1	1

Table C.1 *Continued*

		FIXTURE WIRES					
	Conductor Size	**Metric Designator (Trade Size)**					
Type	**(AWG/ kcmil)**	**16 (½)**	**21 (¾)**	**27 (1)**	**35 (1¼)**	**41 (1½)**	**53 (2)**
FFH-2, RFH-2,	18	8	14	24	41	56	92
RFHH-3	16	7	12	20	34	47	78
SF-2, SFF-2	18	10	18	30	52	71	116
	16	8	15	25	43	58	96
	14	7	12	20	34	47	78
SF-1, SFF-1	18	18	33	53	92	125	206
RFH-1, RFHH-2, TF, TFF, XF, XFF	18	14	24	39	68	92	152
RFHH-2, TF, TFF, XF, XFF	16	11	19	31	55	74	123
XF, XFF	14	8	15	25	43	58	96
TFN, TFFN	18	22	38	63	108	148	244
	16	17	29	48	83	113	186
PF, PFF, PGF, PGFF, PAF, PTF, PTFF, PAFF	18	21	36	59	103	140	231
	16	16	28	46	79	108	179
	14	12	21	34	60	81	134
ZF, ZFF, ZHF, HF, HFF	18	27	47	77	133	181	298
	16	20	35	56	98	133	220
	14	14	25	41	72	98	161
KF-2, KFF-2	18	39	69	111	193	262	433
	16	27	48	78	136	185	305
	14	19	33	54	93	127	209
	12	13	23	37	64	87	144
	10	8	15	25	43	58	96
KF-1, KFF-1	18	46	82	133	230	313	516
	16	33	57	93	161	220	362
	14	22	38	63	108	148	244
	12	14	25	41	72	98	161
	10	9	16	27	47	64	105
XF, XFF	12	4	8	13	23	31	51
	10	3	6	10	18	24	40

Notes:

1. This table is for concentric stranded conductors only. For compact stranded conductors, Table C.1(A) should be used.

2. Two-hour fire-rated RHH cable has ceramifiable insulation which has much larger diameters than other RHH wires. Consult manufacturer's conduit fill tables.

*Types RHH, RHW, and RHW-2 without outer covering.

Table C.1(A) Maximum Number of Compact Conductors in Electrical Metallic Tubing (EMT)
(Based on Table 1, Chapter 9)

		COMPACT CONDUCTORS									
	Conductor Size	Metric Designator (Trade Size)									
Type	(AWG/ kcmil)	16 (½)	21 (¾)	27 (1)	35 (1¼)	41 (1½)	53 (2)	63 (2½)	78 (3)	91 (3½)	103 (4)
THW,	8	2	4	6	11	16	26	46	69	90	115
THW-2,	6	1	3	5	9	12	20	35	53	70	89
THHW	4	1	2	4	6	9	15	26	40	52	67
	2	1	1	3	5	7	11	19	29	38	49
	1	1	1	1	3	4	8	13	21	27	34
	1/0	1	1	1	3	4	7	12	18	23	30
	2/0	0	1	1	2	3	5	10	15	20	25
	3/0	0	1	1	1	3	5	8	13	17	21
	4/0	0	1	1	1	2	4	7	11	14	18
	250	0	0	1	1	1	3	5	8	11	14
	300	0	0	1	1	1	3	5	7	9	12
	350	0	0	1	1	1	2	4	6	8	11
	400	0	0	0	1	1	1	4	6	8	10
	500	0	0	0	1	1	1	3	5	6	8
	600	0	0	0	1	1	1	2	4	5	7
	700	0	0	0	1	1	1	2	3	4	6
	750	0	0	0	0	1	1	1	3	4	5
	900	0	0	0	0	1	1	2	3	4	5
	1000	0	0	0	0	1	1	1	2	3	4
THHN,	8	—	—	—	—	—	—	—	—	—	—
THWN,	6	2	4	7	13	18	29	52	78	102	130
THWN-2	4	1	3	4	8	11	18	32	48	63	81
	2	1	1	3	6	8	13	23	34	45	58
	1	1	1	2	4	6	10	17	26	34	43
	1/0	1	1	1	3	5	8	14	22	29	37
	2/0	1	1	1	3	4	7	12	18	24	30
	3/0	0	1	1	2	3	6	10	15	20	25
	4/0	0	1	1	1	3	5	8	12	16	21
	250	0	1	1	1	1	4	6	10	13	16
	300	0	0	1	1	1	3	5	8	11	14
	350	0	0	1	1	1	3	5	7	10	12
	400	0	0	1	1	1	2	4	6	9	11
	500	0	0	0	1	1	1	4	5	7	9
	600	0	0	0	1	1	1	3	4	6	7
	700	0	0	0	1	1	1	2	4	5	7
	750	0	0	0	1	1	1	2	4	5	6
	900	0	0	0	0	1	1	2	3	4	5
	1000	0	0	0	0	1	1	1	3	3	4
XHHW,	8	3	5	8	15	20	34	59	90	117	149
XHHW-2	6	1	4	6	11	15	25	44	66	87	111
	4	1	3	4	8	11	18	32	48	63	81
	2	1	1	3	6	8	13	23	34	45	58
	1	1	1	2	4	6	10	17	26	34	43
	1/0	1	1	1	3	5	8	14	22	29	37
	2/0	1	1	1	3	4	7	12	18	24	31
	3/0	0	1	1	2	3	6	10	15	20	25
	4/0	0	1	1	1	3	5	8	13	17	21
	250	0	1	1	1	2	4	7	10	13	17
	300	0	0	1	1	1	3	6	9	11	14
	350	0	0	1	1	1	3	5	8	10	13
	400	0	0	1	1	1	2	4	7	9	11
	500	0	0	0	1	1	1	4	6	7	9
	600	0	0	0	1	1	1	3	4	6	8
	700	0	0	0	1	1	1	2	4	5	7
	750	0	0	0	1	1	1	2	3	5	6
	900	0	0	0	0	1	1	2	3	4	5
	1000	0	0	0	0	1	1	1	3	4	5

Definition: *Compact stranding* is the result of a manufacturing process where the standard conductor is compressed to the extent that the interstices (voids between strand wires) are virtually eliminated.

Table C.2 Maximum Number of Conductors or Fixture Wires in Electrical Nonmetallic Tubing (ENT) (*Based on Table 1, Chapter 9*)

		CONDUCTORS					
	Conductor Size	Metric Designator (Trade Size)					
Type	(AWG/ kcmil)	16 (½)	21 (¾)	27 (1)	35 (1¼)	41 (1½)	53 (2)
RHH, RHW, RHW-2	14	3	6	10	19	26	43
	12	2	5	9	16	22	36
	10	1	4	7	13	17	29
	8	1	1	3	6	9	15
	6	1	1	3	5	7	12
	4	1	1	2	4	6	9
	3	1	1	1	3	5	8
	2	0	1	1	3	4	7
	1	0	1	1	1	3	5
	1/0	0	0	1	1	2	4
	2/0	0	0	1	1	1	3
	3/0	0	0	1	1	1	3
	4/0	0	0	1	1	1	2
	250	0	0	0	1	1	1
	300	0	0	0	1	1	1
	350	0	0	0	1	1	1
	400	0	0	0	1	1	1
	500	0	0	0	0	1	1
	600	0	0	0	0	1	1
	700	0	0	0	0	0	1
	750	0	0	0	0	0	1
	800	0	0	0	0	0	1
	900	0	0	0	0	0	1
	1000	0	0	0	0	0	1
	1250	0	0	0	0	0	0
	1500	0	0	0	0	0	0
	1750	0	0	0	0	0	0
	2000	0	0	0	0	0	0
TW	14	7	13	22	40	55	92
	12	5	10	17	31	42	71
	10	4	7	13	23	32	52
	8	1	4	7	13	17	29
RHH*, RHW*, RHW-2*, THHW, THW, THW-2	14	4	8	15	27	37	61
RHH*, RHW*, RHW-2*, THHW, THW	12	3	7	12	21	29	49
	10	3	5	9	17	23	38
RHH*, RHW*, RHW-2*, THHW, THW, THW-2	8	1	3	5	10	14	23

(Continues)

Table C.2 *Continued*

		CONDUCTORS					
	Conductor Size	Metric Designator (Trade Size)					
Type	(AWG/ kcmil)	16 (½)	21 (¾)	27 (1)	35 (1¼)	41 (1½)	53 (2)
RHH*,	6	1	2	4	7	10	17
RHW*,	4	1	1	3	5	8	13
RHW-2*,	3	1	1	2	5	7	11
TW, THW,	2	1	1	2	4	6	9
THHW,	1	0	1	1	3	4	6
THW-2	1/0	0	1	1	2	3	5
	2/0	0	1	1	1	3	5
	3/0	0	0	1	1	2	4
	4/0	0	0	1	1	1	3
	250	0	0	1	1	1	2
	300	0	0	0	1	1	2
	350	0	0	0	1	1	1
	400	0	0	0	1	1	1
	500	0	0	0	1	1	1
	600	0	0	0	0	1	1
	700	0	0	0	0	1	1
	750	0	0	0	0	1	1
	800	0	0	0	0	1	1
	900	0	0	0	0	0	1
	1000	0	0	0	0	0	1
	1250	0	0	0	0	0	1
	1500	0	0	0	0	0	0
	1750	0	0	0	0	0	0
	2000	0	0	0	0	0	0
THHN,	14	10	18	32	58	80	132
THWN,	12	7	13	23	42	58	96
THWN-2	10	4	8	15	26	36	60
	8	2	5	8	15	21	35
	6	1	3	6	11	15	25
	4	1	1	4	7	9	15
	3	1	1	3	5	8	13
	2	1	1	2	5	6	11
	1	1	1	1	3	5	8
	1/0	0	1	1	3	4	7
	2/0	0	1	1	2	3	5
	3/0	0	1	1	1	3	4
	4/0	0	0	1	1	2	4
	250	0	0	1	1	1	3
	300	0	0	1	1	1	2
	350	0	0	0	1	1	2
	400	0	0	0	1	1	1
	500	0	0	0	1	1	1
	600	0	0	0	1	1	1
	700	0	0	0	0	1	1
	750	0	0	0	0	1	1
	800	0	0	0	0	1	1
	900	0	0	0	0	1	1
	1000	0	0	0	0	0	1
FEP, FEPB,	14	10	18	31	56	77	128
PFA, PFAH,	12	7	13	23	41	56	93
TFE	10	5	9	16	29	40	67
	8	3	5	9	17	23	38
	6	1	4	6	12	16	27
	4	1	2	4	8	11	19
	3	1	1	4	7	9	16
	2	1	1	3	5	8	13

Table C.2 *Continued*

		CONDUCTORS					
	Conductor Size	Metric Designator (Trade Size)					
Type	(AWG/ kcmil)	16 (½)	21 (¾)	27 (1)	35 (1¼)	41 (1½)	53 (2)
PFA, PFAH, TFE	1	1	1	1	4	5	9
PFA, PFAH, TFE, Z	1/0	0	1	1	3	4	7
	2/0	0	1	1	2	4	6
	3/0	0	1	1	1	3	5
	4/0	0	1	1	1	2	4
Z	14	12	22	38	68	93	154
	12	8	15	27	48	66	109
	10	5	9	16	29	40	67
	8	3	6	10	18	25	42
	6	1	4	7	13	18	30
	4	1	3	5	9	12	20
	3	1	1	3	6	9	15
	2	1	1	3	5	7	12
	1	1	1	2	4	6	10
XHH, XHHW, XHHW-2, ZW	14	7	13	22	40	55	92
	12	5	10	17	31	42	71
	10	4	7	13	23	32	52
	8	1	4	7	13	17	29
	6	1	3	5	9	13	21
	4	1	1	4	7	9	15
	3	1	1	3	6	8	13
	2	1	1	2	5	6	11
XHH, XHHW, XHHW-2	1	1	1	1	3	5	8
	1/0	0	1	1	3	4	7
	2/0	0	1	1	2	3	6
	3/0	0	1	1	1	3	5
	4/0	0	0	1	1	2	4
	250	0	0	1	1	1	3
	300	0	0	1	1	1	3
	350	0	0	1	1	1	2
	400	0	0	0	1	1	1
	500	0	0	0	1	1	1
	600	0	0	0	1	1	1
	700	0	0	0	0	1	1
	750	0	0	0	0	1	1
	800	0	0	0	0	1	1
	900	0	0	0	0	1	1
	1000	0	0	0	0	0	1
	1250	0	0	0	0	0	1
	1500	0	0	0	0	0	1
	1750	0	0	0	0	0	0
	2000	0	0	0	0	0	0

Table C.2 *Continued*

		FIXTURE WIRES					
	Conductor Size	**Metric Designator (Trade Size)**					
Type	**(AWG/ kcmil)**	**16 (½)**	**21 (¾)**	**27 (1)**	**35 (1¼)**	**41 (1½)**	**53 (2)**
FFH-2,	18	6	12	21	39	53	88
RFH-2,	16	5	10	18	32	45	74
RFHH-3	18	8	15	27	49	67	111
SF-2, SFF-2	16	7	13	22	40	55	92
	14	5	10	18	32	45	74
SF-1, SFF-1	18	15	28	48	86	119	197
RFH-1, RFHH-2, TF, TFF, XF, XFF	18	11	20	35	64	88	145
RFHH-2, TF, TFF, XF, XFF	16	9	16	29	51	71	117
XF, XFF	14	7	13	22	40	55	92
TFN, TFFN	18	18	33	57	102	141	233
	16	13	25	43	78	107	178
PF, PFF, PGF, PGFF, PAF, PTF, PTFF, PAFF	18	17	31	54	97	133	221
	16	13	24	42	75	103	171
	14	10	18	31	56	77	128
ZF, ZFF, ZHF, HF, HFF	18	22	40	70	125	172	285
	16	16	29	51	92	127	210
	14	12	22	38	68	93	154
KF-2, KFF-2	18	31	58	101	182	250	413
	16	22	41	71	128	176	291
	14	15	28	49	88	121	200
	12	10	19	33	60	83	138
	10	7	13	22	40	55	92
KF-1, KFF-1	18	38	69	121	217	298	493
	16	26	49	85	152	209	346
	14	18	33	57	102	141	233
	12	12	22	38	68	93	154
	10	7	14	24	44	61	101
XF, XFF	12	3	7	12	21	29	49
	10	3	5	9	17	23	38

Notes:

1. This table is for concentric stranded conductors only. For compact stranded conductors, Table C.2(A) should be used.

2. Two-hour fire-rated RHH cable has ceramifiable insulation which has much larger diameters than other RHH wires. Consult manufacturer's conduit fill tables.

*Types RHH, RHW, and RHW-2 without outer covering.

Table C.2(A) Maximum Number of Compact Conductors in Electrical Nonmetallic Tubing (ENT) (*Based on Table 1, Chapter 9*)

Type	Conductor Size (AWG/ kcmil)	COMPACT CONDUCTORS					
		Metric Designator (Trade Size)					
		16 (½)	21 (¾)	27 (1)	35 (1¼)	41 (1½)	53 (2)
THW,	8	1	3	6	11	15	25
THW-2,	6	1	2	4	8	11	19
THHW	4	1	1	3	6	8	14
	2	1	1	2	4	6	10
	1	0	1	1	3	4	7
	1/0	0	1	1	3	4	6
	2/0	0	1	1	2	3	5
	3/0	0	1	1	1	3	4
	4/0	0	0	1	1	2	4
	250	0	0	1	1	1	3
	300	0	0	1	1	1	2
	350	0	0	0	1	1	2
	400	0	0	0	1	1	1
	500	0	0	0	1	1	1
	600	0	0	0	1	1	1
	700	0	0	0	0	1	1
	750	0	0	0	0	1	1
	900	0	0	0	0	1	1
	1000	0	0	0	0	0	1
THHN,	8	—	—	—	—	—	—
THWN,	6	1	4	7	12	17	28
THWN-2	4	1	2	4	7	10	17
	2	1	1	3	5	7	12
	1	1	1	2	4	5	9
	1/0	1	1	1	3	5	8
	2/0	0	1	1	3	4	6
	3/0	0	1	1	2	3	5
	4/0	0	1	1	1	2	4
	250	0	0	1	1	1	3
	300	0	0	1	1	1	3
	350	0	0	1	1	1	2
	400	0	0	0	1	1	2
	500	0	0	0	1	1	1
	600	0	0	0	1	1	1
	700	0	0	0	1	1	1
	750	0	0	0	1	1	1
	900	0	0	0	0	1	1
	1000	0	0	0	0	1	1
XHHW,	8	2	4	8	14	19	32
XHHW-2	6	1	3	6	10	14	24
	4	1	2	4	7	10	17
	2	1	1	3	5	7	12
	1	1	1	2	4	5	9
	1/0	1	1	1	3	5	8
	2/0	0	1	1	3	4	7
	3/0	0	1	1	2	3	5
	4/0	0	1	1	1	3	4
	250	0	0	1	1	1	3
	300	0	0	1	1	1	3
	350	0	0	1	1	1	3
	400	0	0	1	1	1	2
	500	0	0	0	1	1	1
	600	0	0	0	1	1	1
	700	0	0	0	1	1	1
	750	0	0	0	1	1	1
	900	0	0	0	0	1	1
	1000	0	0	0	0	1	1

Definition: *Compact stranding* is the result of a manufacturing process where the standard conductor is compressed to the extent that the interstices (voids between strand wires) are virtually eliminated.

Table C.3 Maximum Number of Conductors or Fixture Wires in Flexible Metal Conduit (FMC) (*Based on Table 1, Chapter 9*)

		CONDUCTORS									
	Conductor Size	Metric Designator (Trade Size)									
Type	(AWG/ kcmil)	16 (½)	21 (¾)	27 (1)	35 (1¼)	41 (1½)	53 (2)	63 (2½)	78 (3)	91 (3½)	103 (4)
RHH,	14	4	7	11	17	25	44	67	96	131	171
RHW,	12	3	6	9	14	21	37	55	80	109	142
RHW-2	10	3	5	7	11	17	30	45	64	88	115
	8	1	2	4	6	9	15	23	34	46	60
	6	1	1	3	5	7	12	19	27	37	48
	4	1	1	2	4	5	10	14	21	29	37
	3	1	1	1	3	5	8	13	18	25	33
	2	1	1	1	3	4	7	11	16	22	28
	1	0	1	1	1	2	5	7	10	14	19
	1/0	0	1	1	1	2	4	6	9	12	16
	2/0	0	1	1	1	1	3	5	8	11	14
	3/0	0	0	1	1	1	3	5	7	9	12
	4/0	0	0	1	1	1	2	4	6	8	10
	250	0	0	0	1	1	1	3	4	6	8
	300	0	0	0	1	1	1	2	4	5	7
	350	0	0	0	1	1	1	2	3	5	6
	400	0	0	0	0	1	1	1	3	4	6
	500	0	0	0	0	1	1	1	3	4	5
	600	0	0	0	0	1	1	1	2	3	4
	700	0	0	0	0	0	1	1	1	3	3
	750	0	0	0	0	0	1	1	1	2	3
	800	0	0	0	0	0	1	1	1	2	3
	900	0	0	0	0	0	1	1	1	2	3
	1000	0	0	0	0	0	1	1	1	1	3
	1250	0	0	0	0	0	0	1	1	1	1
	1500	0	0	0	0	0	0	1	1	1	1
	1750	0	0	0	0	0	0	1	1	1	1
	2000	0	0	0	0	0	0	0	1	1	1
TW	14	9	15	23	36	53	94	141	203	277	361
	12	7	11	18	28	41	72	108	156	212	277
	10	5	8	13	21	30	54	81	116	158	207
	8	3	5	7	11	17	30	45	64	88	115
RHH*, RHW*, RHW-2*, THHW, THW, THW-2	14	6	10	15	24	35	62	94	135	184	240
RHH*, RHW*, RHW-2*, THHW, THW	12	5	8	12	19	28	50	75	108	148	193
	10	4	6	10	15	22	39	59	85	115	151
RHH*, RHW*, RHW-2*, THHW, THW, THW-2	8	1	4	6	9	13	23	35	51	69	90

Table C.3 *Continued*

		CONDUCTORS									
	Conductor Size	Metric Designator (Trade Size)									
Type	(AWG/ kcmil)	16 (½)	21 (¾)	27 (1)	35 (1¼)	41 (1½)	53 (2)	63 (2½)	78 (3)	91 (3½)	103 (4)
RHH*,	6	1	3	4	7	10	18	27	39	53	69
RHW*,	4	1	1	3	5	7	13	20	29	39	51
RHW-2*,	3	1	1	3	4	6	11	17	25	34	44
TW,	2	1	1	2	4	5	10	14	21	29	37
THW,	1	1	1	1	2	4	7	10	15	20	26
THHW, THW-2	1/0	0	1	1	1	3	6	9	12	17	22
	2/0	0	1	1	1	3	5	7	10	14	19
	3/0	0	1	1	1	2	4	6	9	12	16
	4/0	0	0	1	1	1	3	5	7	10	13
	250	0	0	1	1	1	3	4	6	8	11
	300	0	0	1	1	1	2	3	5	7	9
	350	0	0	0	1	1	1	3	4	6	8
	400	0	0	0	1	1	1	3	4	6	7
	500	0	0	0	1	1	1	2	3	5	6
	600	0	0	0	0	1	1	1	3	4	5
	700	0	0	0	0	1	1	1	2	3	4
	750	0	0	0	0	1	1	1	2	3	4
	800	0	0	0	0	1	1	1	1	3	4
	900	0	0	0	0	0	1	1	1	3	3
	1000	0	0	0	0	0	1	1	1	2	3
	1250	0	0	0	0	0	1	1	1	1	2
	1500	0	0	0	0	0	0	1	1	1	1
	1750	0	0	0	0	0	0	1	1	1	1
	2000	0	0	0	0	0	0	1	1	1	1
THHN,	14	13	22	33	52	76	134	202	291	396	518
THWN,	12	9	16	24	38	56	98	147	212	289	378
THWN-2	10	6	10	15	24	35	62	93	134	182	238
	8	3	6	9	14	20	35	53	77	105	137
	6	2	4	6	10	14	25	38	55	76	99
	4	1	2	4	6	9	16	24	34	46	61
	3	1	1	3	5	7	13	20	29	39	51
	2	1	1	3	4	6	11	17	24	33	43
	1	1	1	1	3	4	8	12	18	24	32
	1/0	1	1	1	2	4	7	10	15	20	27
	2/0	0	1	1	1	3	6	9	12	17	22
	3/0	0	1	1	1	2	5	7	10	14	18
	4/0	0	1	1	1	1	4	6	8	12	15
	250	0	0	1	1	1	3	5	7	9	12
	300	0	0	1	1	1	3	4	6	8	11
	350	0	0	1	1	1	2	3	5	7	9
	400	0	0	0	1	1	1	3	5	6	8
	500	0	0	0	1	1	1	2	4	5	7
	600	0	0	0	0	1	1	1	3	4	5
	700	0	0	0	0	1	1	1	3	4	5
	750	0	0	0	0	1	1	1	2	3	4
	800	0	0	0	0	1	1	1	2	3	4
	900	0	0	0	0	0	1	1	1	3	4
	1000	0	0	0	0	0	1	1	1	3	3
FEP,	14	12	21	32	51	74	130	196	282	385	502
FEPB,	12	9	15	24	37	54	95	143	206	281	367
PFA,	10	6	11	17	26	39	68	103	148	201	263
PFAH,	8	4	6	10	15	22	39	59	85	115	151
TFE	6	2	4	7	11	16	28	42	60	82	107
	4	1	3	5	7	11	19	29	42	57	75
	3	1	2	4	6	9	16	24	35	48	62
	2	1	1	3	5	7	13	20	29	39	51

(Continues)

Table C.3 *Continued*

		CONDUCTORS								
	Conductor Size	Metric Designator (Trade Size)								
Type	(AWG/ kcmil)	16 (½)	21 (¾)	27 (1)	35 (1¼)	41 (1½)	53 (2)	63 (2½)	78 (3)	91 (3½)	103 (4)
PFA, PFAH, TFE	1	1	1	2	3	5	9	14	20	27	36
PFA, PFAH, TFE, Z	1/0	1	1	1	3	4	8	11	17	23	30
	2/0	1	1	1	2	3	6	9	14	19	24
	3/0	0	1	1	1	3	5	8	11	15	20
	4/0	0	1	1	1	2	4	6	9	13	16
Z	14	15	25	39	61	89	157	236	340	463	605
	12	11	18	28	43	63	111	168	241	329	429
	10	6	11	17	26	39	68	103	148	201	263
	8	4	7	11	17	24	43	65	93	127	166
	6	3	5	7	12	17	30	45	65	89	117
	4	1	3	5	8	12	21	31	45	61	80
	3	1	2	4	6	8	15	23	33	45	58
	2	1	1	3	5	7	12	19	27	37	49
	1	1	1	2	4	6	10	15	22	30	39
XHH, XHHW, XHHW-2, ZW	14	9	15	23	36	53	94	141	203	277	361
	12	7	11	18	28	41	72	108	156	212	277
	10	5	8	13	21	30	54	81	116	158	207
	8	3	5	7	11	17	30	45	64	88	115
	6	1	3	5	8	12	22	33	48	65	85
	4	1	2	4	6	9	16	24	34	47	61
	3	1	1	3	5	7	13	20	29	40	52
	2	1	1	3	4	6	11	17	24	33	44
XHH, XHHW, XHHW-2	1	1	1	1	3	5	8	13	18	25	32
	1/0	1	1	1	2	4	7	10	15	21	27
	2/0	0	1	1	2	3	6	9	13	17	23
	3/0	0	1	1	1	3	5	7	10	14	19
	4/0	0	1	1	1	2	4	6	9	12	15
	250	0	0	1	1	1	3	5	7	10	13
	300	0	0	1	1	1	3	4	6	8	11
	350	0	0	1	1	1	2	4	5	7	9
	400	0	0	0	1	1	1	3	5	6	8
	500	0	0	0	1	1	1	3	4	5	7
	600	0	0	0	0	1	1	1	3	4	5
	700	0	0	0	0	1	1	1	3	4	5
	750	0	0	0	0	1	1	1	2	3	4
	800	0	0	0	0	1	1	1	2	3	4
	900	0	0	0	0	0	1	1	1	3	4
	1000	0	0	0	0	0	1	1	1	3	3
	1250	0	0	0	0	0	1	1	1	1	3
	1500	0	0	0	0	0	1	1	1	1	2
	1750	0	0	0	0	0	0	1	1	1	1
	2000	0	0	0	0	0	0	1	1	1	1

Table C.3 *Continued*

		FIXTURE WIRES					
	Conductor Size	Metric Designator (Trade Size)					
Type	(AWG/ kcmil)	16 (½)	21 (¾)	27 (1)	35 (1¼)	41 (1½)	53 (2)
FFH-2,	18	8	14	22	35	51	90
RFH-2, RFHH-3	16	7	12	19	29	43	76
SF-2, SFF-2	18	11	18	28	44	64	113
	16	9	15	23	36	53	94
	14	7	12	19	29	43	76
SF-1, SFF-1	18	19	32	50	78	114	201
RFH-1, RFHH-2, TF, TFF, XF, XFF	18	14	24	37	58	84	148
RFHH-2, TF, TFF, XF, XFF	16	11	19	30	47	68	120
XF, XFF	14	9	15	23	36	53	94
TFN, TFFN	18	23	38	59	93	135	237
	16	17	29	45	71	103	181
PF, PFF, PGF, PGFF, PAF, PTF, PTFF, PAFF	18	22	36	56	88	128	225
	16	17	28	43	68	99	174
	14	12	21	32	51	74	130
ZF, ZFF, ZHF, HF, HFF	18	28	47	72	113	165	290
	16	20	35	53	83	121	214
	14	15	25	39	61	89	157
KF-2, KFF-2	18	41	68	105	164	239	421
	16	28	48	74	116	168	297
	14	19	33	51	80	116	204
	12	13	23	35	55	80	140
	10	9	15	23	36	53	94
KF-1, KFF-1	18	48	82	125	196	285	503
	16	34	57	88	138	200	353
	14	23	38	59	93	135	237
	12	15	25	39	61	89	157
	10	10	16	25	40	58	103
XF, XFF	12	5	8	12	19	28	50
	10	4	6	10	15	22	39

Notes:

1. This table is for concentric stranded conductors only. For compact stranded conductors, Table C.3(A) should be used.

2. Two-hour fire-rated RHH cable has ceramifiable insulation which has much larger diameters than other RHH wires. Consult manufacturer's conduit fill tables.

*Types RHH, RHW, and RHW-2 without outer covering.

Table C.3(A) Maximum Number of Compact Conductors in Flexible Metal Conduit (FMC)
(Based on Table 1, Chapter 9)

		COMPACT CONDUCTORS									
	Conductor Size	Metric Designator (Trade Size)									
Type	(AWG/ kcmil)	16 (½)	21 (¾)	27 (1)	35 (1¼)	41 (1½)	53 (2)	63 (2½)	78 (3)	91 (3½)	103 (4)
THW,	8	2	4	6	10	14	25	38	55	75	98
THHW,	6	1	3	5	7	11	20	29	43	58	76
THW-2	4	1	2	3	5	8	15	22	32	43	57
	2	1	1	2	4	6	11	16	23	32	42
	1	1	1	1	3	4	7	11	16	22	29
	1/0	1	1	1	2	3	6	10	14	19	25
	2/0	0	1	1	1	3	5	8	12	16	21
	3/0	0	1	1	1	2	4	7	10	14	18
	4/0	0	1	1	1	1	4	6	8	11	15
	250	0	0	1	1	1	3	4	7	9	12
	300	0	0	1	1	1	2	4	6	8	10
	350	0	0	1	1	1	2	3	5	7	9
	400	0	0	0	1	1	1	3	5	6	8
	500	0	0	0	1	1	1	3	4	5	7
	600	0	0	0	0	1	1	1	3	4	6
	700	0	0	0	0	1	1	1	3	4	5
	750	0	0	0	0	1	1	1	2	3	5
	900	0	0	0	0	1	1	1	2	3	4
	1000	0	0	0	0	0	1	1	1	3	4
THHN,	8	—	—	—	—	—	—	—	—	—	—
THWN,	6	3	4	7	11	16	29	43	62	85	111
THWN-2	4	1	3	4	7	10	18	27	38	52	69
	2	1	1	3	5	7	13	19	28	38	49
	1	1	1	2	3	5	9	14	21	28	37
	1/0	1	1	1	3	4	8	12	17	24	31
	2/0	1	1	1	2	4	6	10	14	20	26
	3/0	0	1	1	1	3	5	8	12	17	22
	4/0	0	1	1	1	2	4	7	10	14	18
	250	0	1	1	1	1	3	5	8	11	14
	300	0	0	1	1	1	3	5	7	9	12
	350	0	0	1	1	1	3	4	6	8	10
	400	0	0	1	1	1	2	3	5	7	9
	500	0	0	0	1	1	1	3	4	6	8
	600	0	0	0	1	1	1	2	3	5	6
	700	0	0	0	0	1	1	1	3	4	6
	750	0	0	0	0	1	1	1	3	4	5
	900	0	0	0	0	1	1	1	2	3	4
	1000	0	0	0	0	0	1	1	1	3	4
XHHW,	8	3	5	8	13	19	33	50	71	97	127
XHHW-2	6	2	4	6	9	14	24	37	53	72	95
	4	1	3	4	7	10	18	27	38	52	69
	2	1	1	3	5	7	13	19	28	38	49
	1	1	1	2	3	5	9	14	21	28	37
	1/0	1	1	1	3	4	8	12	17	24	31
	2/0	0	1	1	2	4	7	10	15	20	26
	3/0	0	1	1	1	3	5	8	12	17	22
	4/0	0	1	1	1	2	4	7	10	14	18
	250	0	1	1	1	1	4	5	8	11	14
	300	0	0	1	1	1	3	5	7	9	12
	350	0	0	1	1	1	3	4	6	8	11
	400	0	0	1	1	1	2	4	5	7	10
	500	0	0	0	1	1	1	3	4	6	8
	600	0	0	0	1	1	1	2	3	5	6
	700	0	0	0	0	1	1	1	3	4	6
	750	0	0	0	0	1	1	1	3	4	5
	900	0	0	0	0	1	1	2	2	3	4
	1000	0	0	0	0	1	1	1	2	3	4

Definition: *Compact stranding* is the result of a manufacturing process where the standard conductor is compressed to the extent that the interstices (voids between strand wires) are virtually eliminated.

Table C.4 Maximum Number of Conductors or Fixture Wires in Intermediate Metal Conduit (IMC) (*Based on Table 1, Chapter 9*)

		CONDUCTORS									
	Conductor Size	Metric Designator (Trade Size)									
Type	(AWG/ kcmil)	16 (½)	21 (¾)	27 (1)	35 (1¼)	41 (1½)	53 (2)	63 (2½)	78 (3)	91 (3½)	103 (4)
RHH, RHW, RHW-2	14	4	8	13	22	30	49	70	108	144	186
	12	4	6	11	18	25	41	58	89	120	154
RHH, RHW, RHW-2	10	3	5	8	15	20	33	47	72	97	124
	8	1	3	4	8	10	17	24	38	50	65
	6	1	1	3	6	8	14	19	30	40	52
	4	1	1	3	5	6	11	15	23	31	41
	3	1	1	2	4	6	9	13	21	28	36
	2	1	1	1	3	5	8	11	18	24	31
	1	0	1	1	2	3	5	7	12	16	20
	1/0	0	1	1	1	3	4	6	10	14	18
	2/0	0	1	1	1	2	4	6	9	12	15
	3/0	0	0	1	1	1	3	5	7	10	13
	4/0	0	0	1	1	1	3	4	6	9	11
	250	0	0	1	1	1	1	3	5	6	8
	300	0	0	0	1	1	1	3	4	6	7
	350	0	0	0	1	1	1	2	4	5	7
	400	0	0	0	1	1	1	2	3	5	6
	500	0	0	0	1	1	1	1	3	4	5
	600	0	0	0	0	1	1	1	2	3	4
	700	0	0	0	0	1	1	1	2	3	4
	750	0	0	0	0	1	1	1	1	3	4
	800	0	0	0	0	0	1	1	1	3	3
	900	0	0	0	0	0	1	1	1	2	3
	1000	0	0	0	0	0	1	1	1	2	3
	1250	0	0	0	0	0	1	1	1	1	2
	1500	0	0	0	0	0	0	1	1	1	1
	1750	0	0	0	0	0	0	1	1	1	1
	2000	0	0	0	0	0	0	1	1	1	1
TW	14	10	17	27	47	64	104	147	228	304	392
	12	7	13	21	36	49	80	113	175	234	301
	10	5	9	15	27	36	59	84	130	174	224
	8	3	5	8	15	20	33	47	72	97	124
RHH*, RHW*, RHW-2*, THHW, THW, THW-2	14	6	11	18	31	42	69	98	151	202	261
RHH*, RHW*, RHW-2*, THHW, THW	12	5	9	14	25	34	56	79	122	163	209
	10	4	7	11	19	26	43	61	95	127	163
RHH*, RHW*, RHW-2*, THHW, THW, THW-2	8	2	4	7	12	16	26	37	57	76	98
RHH*, RHW*, RHW-2*, TW, THHW, THW, THW-2	6	1	3	5	9	12	20	28	43	58	75
	4	1	2	4	6	9	15	21	32	43	56

(Continues)

Table C.4 *Continued*

		CONDUCTORS									
	Conductor Size	Metric Designator (Trade Size)									
Type	(AWG/ kcmil)	16 (½)	21 (¾)	27 (1)	35 (1¼)	41 (1½)	53 (2)	63 (2½)	78 (3)	91 (3½)	103 (4)
RHH*,	3	1	1	3	6	8	13	18	28	37	48
RHW*,	2	1	1	3	5	6	11	15	23	31	41
RHW-2*,	1	1	1	1	3	4	7	11	16	22	28
TW,	1/0	1	1	1	3	4	6	9	14	19	24
THW,	2/0	0	1	1	2	3	5	8	12	16	20
THHW,	3/0	0	1	1	1	3	4	6	10	13	17
THW-2	4/0	0	1	1	1	2	4	5	8	11	14
	250	0	0	1	1	1	3	4	7	9	12
	300	0	0	1	1	1	2	4	6	8	10
	350	0	0	1	1	1	2	3	5	7	9
	400	0	0	0	1	1	1	3	4	6	8
	500	0	0	0	1	1	1	2	4	5	7
	600	0	0	0	1	1	1	1	3	4	5
	700	0	0	0	0	1	1	1	3	4	5
	750	0	0	0	0	1	1	1	2	3	4
	800	0	0	0	0	1	1	1	2	3	4
	900	0	0	0	0	1	1	1	2	3	4
	1000	0	0	0	0	0	1	1	1	3	3
	1250	0	0	0	0	0	1	1	1	1	3
	1500	0	0	0	0	0	1	1	1	1	2
	1750	0	0	0	0	0	0	1	1	1	1
	2000	0	0	0	0	0	0	1	1	1	1
THHN,	14	14	24	39	68	91	149	211	326	436	562
THWN,	12	10	17	29	49	67	109	154	238	318	410
THWN-2	10	6	11	18	31	42	68	97	150	200	258
	8	3	6	10	18	24	39	56	86	115	149
	6	2	4	7	13	17	28	40	62	83	107
	4	1	3	4	8	10	17	25	38	51	66
	3	1	2	4	6	9	15	21	32	43	56
	2	1	1	3	5	7	12	17	27	36	47
	1	1	1	2	4	5	9	13	20	27	35
	1/0	1	1	1	3	4	8	11	17	23	29
	2/0	1	1	1	3	4	6	9	14	19	24
	3/0	0	1	1	2	3	5	7	12	16	20
	4/0	0	1	1	1	2	4	6	9	13	17
	250	0	0	1	1	1	3	5	8	10	13
	300	0	0	1	1	1	3	4	7	9	12
	350	0	0	1	1	1	2	4	6	8	10
	400	0	0	1	1	1	2	3	5	7	9
	500	0	0	0	1	1	1	3	4	6	7
	600	0	0	0	1	1	1	2	3	5	6
	700	0	0	0	1	1	1	1	3	4	5
	750	0	0	0	1	1	1	1	3	4	5
	800	0	0	0	0	1	1	1	3	4	5
	900	0	0	0	0	1	1	1	2	3	4
	1000	0	0	0	0	1	1	1	2	3	4
FEP,	14	13	23	38	66	89	145	205	317	423	545
FEPB,	12	10	17	28	48	65	106	150	231	309	398
PFA,	10	7	12	20	34	46	76	107	166	221	285
PFAH,	8	4	7	11	19	26	43	61	95	127	163
TFE	6	3	5	8	14	19	31	44	67	90	116
	4	1	3	5	10	13	21	30	47	63	81
	3	1	3	4	8	11	18	25	39	52	68
	2	1	2	4	6	9	15	21	32	43	56

Table C.4 *Continued*

		CONDUCTORS									
	Conductor Size	Metric Designator (Trade Size)									
Type	(AWG/ kcmil)	16 (½)	21 (¾)	27 (1)	35 (1¼)	41 (1½)	53 (2)	63 (2½)	78 (3)	91 (3½)	103 (4)
PFA, PFAH, TFE	1	1	1	2	4	6	10	14	22	30	39
PFA, PFAH, TFE, Z	1/0	1	1	1	4	5	8	12	19	25	32
	2/0	1	1	1	3	4	7	10	15	21	27
	3/0	0	1	1	2	3	6	8	13	17	22
	4/0	0	1	1	1	3	5	7	10	14	18
Z	14	16	28	46	79	107	175	247	381	510	657
	12	11	20	32	56	76	124	175	271	362	466
	10	7	12	20	34	46	76	107	166	221	285
	8	4	7	12	21	29	48	68	105	140	180
	6	3	5	9	15	20	33	47	73	98	127
	4	1	3	6	10	14	23	33	50	67	87
	3	1	2	4	7	10	17	24	37	49	63
	2	1	1	3	6	8	14	20	30	41	53
	1	1	1	3	5	7	11	16	25	33	43
XHH, XHHW, XHHW-2, ZW	14	10	17	27	47	64	104	147	228	304	392
	12	7	13	21	36	49	80	113	175	234	301
	10	5	9	15	27	36	59	84	130	174	224
	8	3	5	8	15	20	33	47	72	97	124
	6	1	4	6	11	15	24	35	53	71	92
	4	1	3	4	8	11	18	25	39	52	67
	3	1	2	4	7	9	15	21	33	44	56
	2	1	1	3	5	7	12	18	27	37	47
XHH, XHHW, XHHW-2	1	1	1	2	4	5	9	13	20	27	35
	1/0	1	1	1	3	5	8	11	17	23	30
	2/0	1	1	1	3	4	6	9	14	19	25
	3/0	0	1	1	2	3	5	7	12	16	20
	4/0	0	1	1	1	2	4	6	10	13	17
	250	0	0	1	1	1	3	5	8	11	14
	300	0	0	1	1	1	3	4	7	9	12
	350	0	0	1	1	1	3	4	6	8	10
	400	0	0	1	1	1	2	3	5	7	9
	500	0	0	0	1	1	1	3	4	6	8
	600	0	0	0	1	1	1	2	3	5	6
	700	0	0	0	1	1	1	1	3	4	5
	750	0	0	0	1	1	1	1	3	4	5
	800	0	0	0	0	1	1	1	3	4	5
	900	0	0	0	0	1	1	1	2	3	4
	1000	0	0	0	0	1	1	1	2	3	4
	1250	0	0	0	0	0	1	1	1	2	3
	1500	0	0	0	0	0	1	1	1	1	2
	1750	0	0	0	0	0	1	1	1	1	2
	2000	0	0	0	0	0	0	1	1	1	1

(Continues)

Table C.4 *Continued*

	Conductor Size (AWG/ kcmil)	Metric Designator (Trade Size)					
Type		16 (½)	21 (¾)	27 (1)	35 (1¼)	41 (1½)	53 (2)
FHH-2, RFH-2, RFHH-3	18	9	16	26	45	61	100
	16	8	13	22	38	51	84
SF-2, SFF-2	18	12	20	33	57	77	126
	16	10	17	27	47	64	104
	14	8	13	22	38	51	84
SF-1, SFF-1	18	21	36	59	101	137	223
RFH-1, RFHH-2, TF, TFF, XF, XFF	18	15	26	43	75	101	165
RFH-2, TF, TFF, XF, XFF	16	12	21	35	60	81	133
XF, XFF	14	10	17	27	47	64	104
TFN, TFFN	18	25	42	69	119	161	264
	16	19	32	53	91	123	201
PF, PFF, PGF, PGFF, PAF, PTF, PTFF, PAFF	18	23	40	66	113	153	250
	16	18	31	51	87	118	193
	14	13	23	38	66	89	145
ZF, ZFF, ZHF, HF, HFF	18	30	52	85	146	197	322
	16	22	38	63	108	145	238
	14	16	28	46	79	107	175
KF-2, KFF-2	18	44	75	123	212	287	468
	16	31	53	87	149	202	330
	14	21	36	60	103	139	227
	12	14	25	41	70	95	156
	10	10	17	27	47	64	104
KF-1, KFF-1	18	52	90	147	253	342	558
	16	37	63	103	178	240	392
	14	25	42	69	119	161	264
	12	16	28	46	79	107	175
	10	10	18	30	52	70	114
XF, XFF	12	5	9	14	25	34	56
	10	4	7	11	19	26	43

Notes:

1. This table is for concentric stranded conductors only. For compact stranded conductors, Table C.4(A) should be used.

2. Two-hour fire-rated RHH cable has ceramifiable insulation which has much larger diameters than other RHH wires. Consult manufacturer's conduit fill tables.

*Types RHH, RHW, and RHW-2 without outer covering.

Table C.4(A) Maximum Number of Compact Conductors in Intermediate Metal Conduit (IMC) *(Based on Table 1, Chapter 9)*

	Conductor Size (AWG/ kcmil)	COMPACT CONDUCTORS									
		Metric Designator (Trade Size)									
Type		16 (½)	21 (¾)	27 (1)	35 (1¼)	41 (1½)	53 (2)	63 (2½)	78 (3)	91 (3½)	103 (4)
THW, THW-2, THHW	8	2	4	7	13	17	28	40	62	83	107
	6	1	3	6	10	13	22	31	48	64	82
	4	1	2	4	7	10	16	23	36	48	62
	2	1	1	3	5	7	12	17	26	35	45
	1	1	1	1	4	5	8	12	18	25	32
	1/0	1	1	1	3	4	7	10	16	21	27
	2/0	0	1	1	3	4	6	9	13	18	23
	3/0	0	1	1	2	3	5	7	11	15	20
	4/0	0	1	1	1	2	4	6	9	13	16
	250	0	0	1	1	1	3	5	7	10	13
	300	0	0	1	1	1	3	4	6	9	11
	350	0	0	1	1	1	2	4	6	8	10
	400	0	0	1	1	1	2	3	5	7	9
	500	0	0	0	1	1	1	3	4	6	8
	600	0	0	0	1	1	1	2	3	5	6
	700	0	0	0	1	1	1	1	3	4	5
	750	0	0	0	1	1	1	1	3	4	5
	900	0	0	0	0	1	1	1	2	3	4
	1000	0	0	0	0	1	1	1	2	3	4
THHN, THWN, THWN-2	8	—	—	—	—	—	—	—	—	—	—
	6	3	5	8	14	19	32	45	70	93	120
	4	1	3	5	9	12	20	28	43	58	74
	2	1	1	3	6	8	14	20	31	41	53
	1	1	1	3	5	6	10	15	23	31	40
	1/0	1	1	2	4	5	9	13	20	26	34
	2/0	1	1	1	3	4	7	10	16	22	28
	3/0	0	1	1	3	4	6	9	14	18	24
	4/0	0	1	1	2	3	5	7	11	15	19
	250	0	1	1	1	2	4	6	9	12	15
	300	0	0	1	1	1	3	5	7	10	13
	350	0	0	1	1	1	3	4	7	9	11
	400	0	0	1	1	1	2	4	6	8	10
	500	0	0	1	1	1	2	3	5	7	9
	600	0	0	0	1	1	1	2	4	5	7
	700	0	0	0	1	1	1	2	3	5	6
	750	0	0	0	1	1	1	1	3	4	6
	900	0	0	0	0	1	1	2	3	3	5
	1000	0	0	0	0	1	1	1	2	3	4
XHHW, XHHW-2	8	3	6	9	16	22	37	52	80	107	138
	6	2	4	7	12	16	27	38	59	80	103
	4	1	3	5	9	12	20	28	43	58	74
	2	1	1	3	6	8	14	20	31	41	53
	1	1	1	3	5	6	10	15	23	31	40
	1/0	1	1	2	4	5	9	13	20	26	34
	2/0	1	1	1	3	4	7	11	17	22	29
	3/0	0	1	1	3	4	6	9	14	18	24
	4/0	0	1	1	2	3	5	7	11	15	20
	250	0	1	1	1	2	4	6	9	12	16
	300	0	0	1	1	1	3	5	8	10	13
	350	0	0	1	1	1	3	4	7	9	12
	400	0	0	1	1	1	3	4	6	8	11
	500	0	0	1	1	1	2	3	5	7	9
	600	0	0	0	1	1	1	2	4	5	7
	700	0	0	0	1	1	1	2	3	5	6
	750	0	0	0	1	1	1	1	3	4	6
	900	0	0	0	0	1	1	2	3	4	5
	1000	0	0	0	0	1	1	1	2	3	4

Definition: *Compact stranding* is the result of a manufacturing process where the standard conductor is compressed to the extent that interstices (voids between strand wires) are virtually eliminated.

Table C.5 Maximum Number of Conductors or Fixture Wires in Liquidtight Flexible Nonmetallic Conduit (Type LFNC-B*) (*Based on Table 1, Chapter 9*)

		CONDUCTORS						
	Conductor Size	Metric Designator (Trade Size)						
Type	(AWG/ kcmil)	12 (⅜)	16 (½)	21 (¾)	27 (1)	35 (1¼)	41 (1½)	53 (2)
RHH,	14	2	4	7	12	21	27	44
RHW,	12	1	3	6	10	17	22	36
RHW-2	10	1	3	5	8	14	18	29
	8	1	1	2	4	7	9	15
	6	1	1	1	3	6	7	12
	4	0	1	1	2	4	6	9
	3	0	1	1	1	4	5	8
	2	0	1	1	1	3	4	7
	1	0	0	1	1	1	3	5
	1/0	0	0	1	1	1	2	4
	2/0	0	0	1	1	1	1	3
	3/0	0	0	0	1	1	1	3
	4/0	0	0	0	1	1	1	2
	250	0	0	0	0	1	1	1
	300	0	0	0	0	1	1	1
	350	0	0	0	0	1	1	1
	400	0	0	0	0	1	1	1
	500	0	0	0	0	1	1	1
	600	0	0	0	0	0	1	1
	700	0	0	0	0	0	0	1
	750	0	0	0	0	0	0	1
	800	0	0	0	0	0	0	1
	900	0	0	0	0	0	0	1
	1000	0	0	0	0	0	0	1
	1250	0	0	0	0	0	0	0
	1500	0	0	0	0	0	0	0
	1750	0	0	0	0	0	0	0
	2000	0	0	0	0	0	0	0
TW	14	5	9	15	25	44	57	93
	12	4	7	12	19	33	43	71
	10	3	5	9	14	25	32	53
	8	1	3	5	8	14	18	29
RHH[†], RHW[†], RHW-2[†], THHW, THW, THW-2	14	3	6	10	16	29	38	62
RHH[†], RHW[†], RHW-2[†], THHW, THW	12	3	5	8	13	23	30	50
	10	1	3	6	10	18	23	39
RHH[†], RHW[†], RHW-2[†], THHW, THW, THW-2	8	1	1	4	6	11	14	23
RHH[†], RHW[†], RHW-2[†], TW, THW, THHW, THW-2	6	1	1	3	5	8	11	18
	4	1	1	1	3	6	8	13
	3	1	1	1	3	5	7	11

Table C.5 *Continued*

		CONDUCTORS						
	Conductor Size	Metric Designator (Trade Size)						
Type	(AWG/ kcmil)	12 (⅜)	16 (½)	21 (¾)	27 (1)	35 (1¼)	41 (1½)	53 (2)
RHH[†], RHW[†], RHW-2[†], TW, THW, THHW, THW-2	2	0	1	1	2	4	6	9
	1	0	1	1	1	3	4	7
	1/0	0	0	1	1	2	3	6
	2/0	0	0	1	1	2	3	5
	3/0	0	0	1	1	1	2	4
	4/0	0	0	0	1	1	1	3
	250	0	0	0	1	1	1	3
	300	0	0	0	1	1	1	2
	350	0	0	0	0	1	1	1
	400	0	0	0	0	1	1	1
	500	0	0	0	0	1	1	1
	600	0	0	0	0	1	1	1
	700	0	0	0	0	0	1	1
	750	0	0	0	0	0	1	1
	800	0	0	0	0	0	1	1
	900	0	0	0	0	0	0	1
	1000	0	0	0	0	0	0	1
	1250	0	0	0	0	0	0	1
	1500	0	0	0	0	0	0	0
	1750	0	0	0	0	0	0	0
	2000	0	0	0	0	0	0	0
THHN, THWN, THWN-2	14	8	13	22	36	63	81	133
	12	5	9	16	26	46	59	97
	10	3	6	10	16	29	37	61
	8	1	3	6	9	16	21	35
	6	1	2	4	7	12	15	25
	4	1	1	2	4	7	9	15
	3	1	1	1	3	6	8	13
	2	1	1	1	3	5	7	11
	1	0	1	1	1	4	5	8
	1/0	0	1	1	1	3	4	7
	2/0	0	0	1	1	2	3	6
	3/0	0	0	1	1	1	3	5
	4/0	0	0	1	1	1	2	4
	250	0	0	0	1	1	1	3
	300	0	0	0	1	1	1	3
	350	0	0	0	1	1	1	2
	400	0	0	0	0	1	1	1
	500	0	0	0	0	1	1	1
	600	0	0	0	0	1	1	1
	700	0	0	0	0	1	1	1
	750	0	0	0	0	0	1	1
	800	0	0	0	0	0	1	1
	900	0	0	0	0	0	1	1
	1000	0	0	0	0	0	0	1
FEP, FEPB, PFA, PFAH, TFE	14	7	12	21	35	61	79	129
	12	5	9	15	25	44	57	94
	10	4	6	11	18	32	41	68
	8	1	3	6	10	18	23	39
	6	1	2	4	7	13	17	27
	4	1	1	3	5	9	12	19
	3	1	1	2	4	7	10	16
	2	1	1	1	3	6	8	13
PFA, PFAH, TFE	1	0	1	1	2	4	5	9
PFA, PFAH	1/0	0	1	1	1	3	4	7

(Continues)

Table C.5 *Continued*

	CONDUCTORS						
	Conductor Size	Metric Designator (Trade Size)					
Type	(AWG/ kcmil)	12 (⅜)	16 (½)	21 (¾)	27 (1)	35 (1¼)	41 (1½)	53 (2)
TFE, Z	2/0	0	1	1	1	3	4	6
	3/0	0	0	1	1	2	3	5
	4/0	0	0	1	1	1	2	4
Z	14	9	15	26	42	73	95	156
	12	6	10	18	30	52	67	111
	10	4	6	11	18	32	41	68
	8	2	4	7	11	20	26	43
	6	1	3	5	8	14	18	30
	4	1	1	3	5	9	12	20
	3	1	1	2	4	7	9	15
	2	0	1	1	3	6	7	12
	1	0	1	1	2	5	6	10
XHH, XHHW, XHHW-2, ZW	14	5	9	15	25	44	57	93
	12	4	7	12	19	33	43	71
	10	3	5	9	14	25	32	53
	8	1	3	5	8	14	18	29
	6	1	1	3	6	10	13	22
	4	1	1	2	4	7	9	16
	3	1	1	1	3	6	8	13
	2	1	1	1	3	5	7	11
XHH, XHHW, XHHW-2	1	0	1	1	1	4	5	8
	1/0	0	1	1	1	3	4	7
	2/0	0	0	1	1	2	3	6
	3/0	0	0	1	1	1	3	5
	4/0	0	0	1	1	1	2	4
	250	0	0	0	1	1	1	3
	300	0	0	0	1	1	1	3
	350	0	0	0	1	1	1	2
	400	0	0	0	0	1	1	1
	500	0	0	0	0	1	1	1
	600	0	0	0	0	1	1	1
	700	0	0	0	0	1	1	1
	750	0	0	0	0	0	1	1
	800	0	0	0	0	0	1	1
	900	0	0	0	0	0	1	1
	1000	0	0	0	0	0	0	1
	1250	0	0	0	0	0	0	1
	1500	0	0	0	0	0	0	1
	1750	0	0	0	0	0	0	0
	2000	0	0	0	0	0	0	0
	FIXTURE WIRES						
FFH-2, RFH-2	18	5	8	15	24	42	54	89
	16	4	7	12	20	35	46	75
SF-2, SFF-2	18	6	11	19	30	53	69	113
	16	5	9	15	25	44	57	93
	14	4	7	12	20	35	46	75
SF-1, SFF-1	18	11	19	33	53	94	122	199
RFH-1, RFHH-2, TF, TFF, XF, XFF	18	8	14	24	39	69	90	147
RFHH-2, TF, TFF, XF, XFF	16	7	11	20	32	56	72	119
XF, XFF	14	5	9	15	25	44	57	93
TFN, TFFN	18	14	23	39	63	111	144	236
	16	10	17	30	48	85	110	180

Table C.5 *Continued*

	Conductor Size (AWG/ kcmil)	FIXTURE WIRES Metric Designator (Trade Size)						
Type		12 (⅜)	16 (½)	21 (¾)	27 (1)	35 (1¼)	41 (1½)	53 (2)
PF, PFF, PGF, PGFF, PAF, PTF, PTFF, PAFF	18	13	21	37	60	105	136	223
	16	10	16	29	46	81	105	173
	14	7	12	21	35	61	79	129
HF, HFF, ZF, ZFF, ZHF	18	17	28	48	77	136	176	288
	16	12	20	35	57	100	129	212
	14	9	15	26	42	73	95	156
KF-2, KFF-2	18	24	40	70	112	197	255	418
	16	17	28	49	79	139	180	295
	14	12	19	34	54	95	123	202
	12	8	13	23	37	65	85	139
	10	5	9	15	25	44	57	93
KF-1, KFF-1	18	29	48	83	134	235	304	499
	16	20	34	58	94	165	214	350
	14	14	23	39	63	111	144	236
	12	9	15	26	42	73	95	156
	10	6	10	17	27	48	62	102
XF, XFF	12	3	5	8	13	23	30	50
	10	1	3	6	10	18	23	39

Notes:

1. This table is for concentric stranded conductors only. For compact stranded conductors, Table C.5(A) should be used.

2. Two-hour fire-rated RHH cable has ceramifiable insulation which has much larger diameters than other RHH wires. Consult manufacturer's conduit fill tables.

*Corresponds to 356.2(2).

†Types RHH, RHW, and RHW-2 without outer covering.

Table C.5(A) Maximum Number of Compact Conductors in Liquidtight Flexible Nonmetallic Conduit (Type LFNC-B*) (*Based on Table 1, Chapter 9*)

		COMPACT CONDUCTORS						
	Conductor Size	Metric Designator (Trade Size)						
Type	(AWG/kcmil)	12 (⅜)	16 (½)	21 (¾)	27 (1)	35 (1¼)	41 (1½)	53 (2)
THW,	8	1	2	4	7	12	15	25
THW-2,	6	1	1	3	5	9	12	19
THHW	4	1	1	2	4	7	9	14
	2	1	1	1	3	5	6	11
	1	0	1	1	1	3	4	7
	1/0	0	1	1	1	3	4	6
	2/0	0	0	1	1	2	3	5
	3/0	0	0	1	1	1	3	4
	4/0	0	0	1	1	1	2	4
	250	0	0	0	1	1	1	3
	300	0	0	0	1	1	1	2
	350	0	0	0	1	1	1	2
	400	0	0	0	0	1	1	1
	500	0	0	0	0	1	1	1
	600	0	0	0	0	1	1	1
	700	0	0	0	0	1	1	1
	750	0	0	0	0	0	1	1
	900	—	0	0	0	0	1	1
	1000	0	0	0	0	0	1	1
THHN,	8	—	—	—	—	—	—	—
THWN,	6	1	2	4	7	13	17	28
THWN-2	4	1	1	3	4	8	11	17
	2	1	1	1	3	6	7	12
	1	0	1	1	2	4	6	9
	1/0	0	1	1	1	4	5	8
	2/0	0	1	1	1	3	4	6
	3/0	0	0	1	1	2	3	5
	4/0	0	0	1	1	1	3	4
	250	0	0	1	1	1	1	3
	300	0	0	0	1	1	1	3
	350	0	0	0	1	1	1	2
	400	0	0	0	1	1	1	2
	500	0	0	0	0	1	1	1
	600	0	0	0	0	1	1	1
	700	0	0	0	0	1	1	1
	750	0	0	0	0	1	1	1
	900	—	0	0	0	0	1	1
	1000	0	0	0	0	0	1	1
XHHW,	8	1	3	5	9	15	20	33
XHHW-2	6	1	2	4	6	11	15	24
	4	1	1	3	4	8	11	17
	2	1	1	1	3	6	7	12
	1	0	1	1	2	4	6	9
	1/0	0	1	1	1	4	5	8
	2/0	0	1	1	1	3	4	7
	3/0	0	0	1	1	2	3	5
	4/0	0	0	1	1	1	3	4
	250	0	0	1	1	1	1	3
	300	0	0	0	1	1	1	3
	350	0	0	0	1	1	1	3
	400	0	0	0	1	1	1	2
	500	0	0	0	0	1	1	1
	600	0	0	0	0	1	1	1
	700	0	0	0	0	1	1	1
	750	0	0	0	0	1	1	1
	900	—	0	0	0	0	1	1
	1000	0	0	0	0	0	1	1

*Corresponds to 356.2(2).

Definition: *Compact stranding* is the result of a manufacturing process where the standard conductor is compressed to the extent that the interstices (voids between strand wires) are virtually eliminated.

Table C.6 Maximum Number of Conductors or Fixture Wires in Liquidtight Flexible Nonmetallic Conduit (Type LFNC-A*) (*Based on Table 1, Chapter 9*)

		CONDUCTORS						
	Conductor Size	Metric Designator (Trade Size)						
Type	(AWG/ kcmil)	12 (³⁄₈)	16 (½)	21 (¾)	27 (1)	35 (1¼)	41 (1½)	53 (2)
RHH, RHW, RHW-2	14	2	4	7	11	20	27	45
	12	1	3	6	9	17	23	38
	10	1	3	5	8	13	18	30
	8	1	1	2	4	7	9	16
	6	1	1	1	3	5	7	13
	4	0	1	1	2	4	6	10
	3	0	1	1	1	4	5	8
	2	0	1	1	1	3	4	7
	1	0	0	1	1	1	3	5
	1/0	0	0	1	1	1	2	4
	2/0	0	0	1	1	1	1	4
	3/0	0	0	0	1	1	1	3
	4/0	0	0	0	1	1	1	3
	250	0	0	0	0	1	1	1
	300	0	0	0	0	1	1	1
	350	0	0	0	0	1	1	1
	400	0	0	0	0	1	1	1
	500	0	0	0	0	0	1	1
	600	0	0	0	0	0	1	1
	700	0	0	0	0	0	0	1
	750	0	0	0	0	0	0	1
	800	0	0	0	0	0	0	1
	900	0	0	0	0	0	0	1
	1000	0	0	0	0	0	0	1
	1250	0	0	0	0	0	0	0
	1500	0	0	0	0	0	0	0
	1750	0	0	0	0	0	0	0
	2000	0	0	0	0	0	0	0
TW	14	5	9	15	24	43	58	96
	12	4	7	12	19	33	44	74
	10	3	5	9	14	24	33	55
	8	1	3	5	8	13	18	30
RHH†, RHW†, RHW-2†, THHW, THW, THW-2	14	3	6	10	16	28	38	64
RHH†, RHW†, RHW-2†, THHW, THW†	12	3	4	8	13	23	31	51
	10	1	3	6	10	18	24	40
RHH†, RHW†, RHW-2†, THHW, THW, THW-2	8	1	1	4	6	10	14	24

(Continues)

Table C.6 *Continued*

		CONDUCTORS						
	Conductor Size	Metric Designator (Trade Size)						
Type	(AWG/ kcmil)	12 (⅜)	16 (½)	21 (¾)	27 (1)	35 (1¼)	41 (1½)	53 (2)
RHH†,	6	1	1	3	4	8	11	18
RHW†,	4	1	1	1	3	6	8	13
RHW-2†,	3	1	1	1	3	5	7	11
TW,	2	0	1	1	2	4	6	10
THW,	1	0	1	1	1	3	4	7
THHW, THW-2	1/0	0	0	1	1	2	3	6
	2/0	0	0	1	1	1	3	5
	3/0	0	0	1	1	1	2	4
	4/0	0	0	0	1	1	1	3
	250	0	0	0	1	1	1	3
	300	0	0	0	1	1	1	2
	350	0	0	0	0	1	1	1
	400	0	0	0	0	1	1	1
	500	0	0	0	0	1	1	1
	600	0	0	0	0	1	1	1
	700	0	0	0	0	0	1	1
	750	0	0	0	0	0	1	1
	800	0	0	0	0	0	1	1
	900	0	0	0	0	0	0	1
	1000	0	0	0	0	0	0	1
	1250	0	0	0	0	0	0	1
	1500	0	0	0	0	0	0	1
	1750	0	0	0	0	0	0	0
	2000	0	0	0	0	0	0	0
THHN,	14	8	13	22	35	62	83	137
THWN,	12	5	9	16	25	45	60	100
THWN-2	10	3	6	10	16	28	38	63
	8	1	3	6	9	16	22	36
	6	1	2	4	6	12	16	26
	4	1	1	2	4	7	9	16
	3	1	1	1	3	6	8	13
	2	1	1	1	3	5	7	11
	1	0	1	1	1	4	5	8
	1/0	0	1	1	1	3	4	7
	2/0	0	0	1	1	2	3	6
	3/0	0	0	1	1	1	3	5
	4/0	0	0	1	1	1	2	4
	250	0	0	0	1	1	1	3
	300	0	0	0	1	1	1	3
	350	0	0	0	1	1	1	2
	400	0	0	0	0	1	1	1
	500	0	0	0	0	1	1	1
	600	0	0	0	0	1	1	1
	700	0	0	0	0	1	1	1
	750	0	0	0	0	0	1	1
	800	0	0	0	0	0	1	1
	900	0	0	0	0	0	1	1
	1000	0	0	0	0	0	0	1
FEP,	14	7	12	21	34	60	80	133
FEPB,	12	5	9	15	25	44	59	97
PFA,	10	4	6	11	18	31	42	70
PFAH,	8	1	3	6	10	18	24	40
TFE	6	1	2	4	7	13	17	28
	4	1	1	3	5	9	12	20
	3	1	1	2	4	7	10	16
	2	1	1	1	3	6	8	13

Table C.6 *Continued*

	Conductor Size (AWG/ kcmil)	**CONDUCTORS**						
		Metric Designator (Trade Size)						
Type		**12** (⅜)	**16** (½)	**21** (¾)	**27** (1)	**35** (1¼)	**41** (1½)	**53** (2)
PFA, PFAH, TFE	1	0	1	1	2	4	5	9
PFA, PFAH, TFE, Z	1/0	0	1	1	1	3	5	8
	2/0	0	1	1	1	3	4	6
	3/0	0	0	1	1	2	3	5
	4/0	0	0	1	1	1	2	4
Z	14	9	15	25	41	72	97	161
	12	6	10	18	29	51	69	114
	10	4	6	11	18	31	42	70
	8	2	4	7	11	20	26	44
	6	1	3	5	8	14	18	31
	4	1	1	3	5	9	13	21
	3	1	1	2	4	7	9	15
	2	1	1	1	3	6	8	13
	1	1	1	1	2	4	6	10
XHH, XHHW, XHHW-2, ZW	14	5	9	15	24	43	58	96
	12	4	7	12	19	33	44	74
	10	3	5	9	14	24	33	55
	8	1	3	5	8	13	18	30
	6	1	1	3	5	10	13	22
	4	1	1	2	4	7	10	16
	3	1	1	1	3	6	8	14
	2	1	1	1	3	5	7	11
XHH, XHHW, XHHW-2	1	0	1	1	1	4	5	8
	1/0	0	1	1	1	3	4	7
	2/0	0	0	1	1	2	3	6
	3/0	0	0	1	1	1	3	5
	4/0	0	0	1	1	1	2	4
	250	0	0	0	1	1	1	3
	300	0	0	0	1	1	1	3
	350	0	0	0	1	1	1	2
	400	0	0	0	0	1	1	1
	500	0	0	0	0	1	1	1
	600	0	0	0	0	1	1	1
	700	0	0	0	0	1	1	1
	750	0	0	0	0	0	1	1
	800	0	0	0	0	0	1	1
	900	0	0	0	0	0	1	1
	1000	0	0	0	0	0	0	1
	1250	0	0	0	0	0	0	1
	1500	0	0	0	0	0	0	1
	1750	0	0	0	0	0	0	0
	2000	0	0	0	0	0	0	0

(Continues)

Table C.6 *Continued*

	FIXTURE WIRES						
	Conductor Size	Metric Designator (Trade Size)					
Type	(AWG/ kcmil)	12 (⅜)	16 (½)	21 (¾)	27 (1)	35 (1¼)	41 (1½)	53 (2)
FFH-2, RFH-2, RFHH-3	18	5	8	14	23	41	55	92
	16	4	7	12	20	35	47	77
SF-2, SFF-2	18	6	11	18	29	52	70	116
	16	5	9	15	24	43	58	96
	14	4	7	12	20	35	47	77
SF-1, SFF-1	18	12	19	33	52	92	124	205
RFH-1, RFHH-2, TF, TFF, XF, XFF	18	8	14	24	39	68	91	152
RFHH-2, TF, TFF, XF, XFF	16	7	11	19	31	55	74	122
XF, XFF	14	5	9	15	24	43	58	96
TFN, TFFN	18	14	22	39	62	109	146	243
	16	10	17	29	47	83	112	185
PF, PFF, PGF, PGFF, PAF, PTF, PTFF, PAFF	18	13	21	37	59	103	139	230
	16	10	16	28	45	80	107	178
	14	7	12	21	34	60	80	133
HF, HFF, ZF, ZFF, ZHF	18	17	27	47	76	133	179	297
	16	12	20	35	56	98	132	219
	14	9	15	25	41	72	97	161
KF-2, KFF-2	18	25	40	69	110	193	260	431
	16	17	28	48	77	136	183	303
	14	12	19	33	53	94	126	209
	12	8	13	23	36	64	86	143
	10	5	9	15	24	43	58	96
KF-1, KFF-1	18	29	48	82	131	231	310	514
	16	21	33	57	92	162	218	361
	14	14	22	39	62	109	146	243
	12	9	15	25	41	72	97	161
	10	6	10	17	27	47	63	105
XF, XFF	12	3	4	8	13	23	31	51
	10	1	3	6	10	18	24	40

Notes:

1. This table is for concentric stranded conductors only. For compact stranded conductors, Table C.6(A) should be used.

2. Two-hour fire-rated RHH cable has ceramifiable insulation which has much larger diameters than other RHH wires. Consult manufacturer's conduit fill tables.

*Corresponds to 356.2(1).

⁺Types RHH, RHW, and RHW-2 without outer covering.

Table C.6(A) Maximum Number of Compact Conductors in Liquidtight Flexible Nonmetallic Conduit (Type LFNC-A*) (*Based on Table 1, Chapter 9*)

		COMPACT CONDUCTORS						
	Conductor Size	Metric Designator (Trade Size)						
Type	(AWG/ kcmil)	12 (⅜)	16 (½)	21 (¾)	27 (1)	35 (1¼)	41 (1½)	53 (2)
THW,	8	1	2	4	6	11	16	26
THW-2,	6	1	1	3	5	9	12	20
THHW	4	1	1	2	4	7	9	15
	2	1	1	1	3	5	6	11
	1	0	1	1	1	3	4	8
	1/0	0	1	1	1	3	4	7
	2/0	0	0	1	1	2	3	5
	3/0	0	0	1	1	1	3	5
	4/0	0	0	1	1	1	2	4
	250	0	0	0	1	1	1	3
	300	0	0	0	1	1	1	3
	350	0	0	0	1	1	1	2
	400	0	0	0	0	1	1	1
	500	0	0	0	0	1	1	1
	600	0	0	0	0	1	1	1
	700	0	0	0	0	1	1	1
	750	0	0	0	0	0	1	1
	900	—	0	0	0	0	1	1
	1000	0	0	0	0	0	1	1
THHN,	8	—	—	—	—	—	—	—
THWN,	6	1	2	4	7	13	18	29
THWN-2	4	1	1	3	4	8	11	18
	2	1	1	1	3	6	8	13
	1	0	1	1	2	4	6	10
	1/0	0	1	1	1	3	5	8
	2/0	0	1	1	1	3	4	7
	3/0	0	0	1	1	2	3	6
	4/0	0	0	1	1	1	3	5
	250	0	0	1	1	1	1	3
	300	0	0	0	1	1	1	3
	350	0	0	0	1	1	1	3
	400	0	0	0	1	1	1	2
	500	0	0	0	0	1	1	1
	600	0	0	0	0	1	1	1
	700	0	0	0	0	1	1	1
	750	0	0	0	0	1	1	1
	900	—	0	0	0	0	1	1
	1000	0	0	0	0	0	1	1
XHHW,	8	1	3	5	8	15	20	34
XHHW-2	6	1	2	4	6	11	15	25
	4	1	1	3	4	8	11	18
	2	1	1	1	3	6	8	13
	1	0	1	1	2	4	6	10
	1/0	0	1	1	1	3	5	8
	2/0	0	1	1	1	3	4	7
	3/0	0	0	1	1	2	3	6
	4/0	0	0	1	1	1	3	5
	250	0	0	1	1	1	2	4
	300	0	0	0	1	1	1	3
	350	0	0	0	1	1	1	3
	400	0	0	0	1	1	1	2
	500	0	0	0	0	1	1	1
	600	0	0	0	0	1	1	1
	700	0	0	0	0	1	1	1
	750	0	0	0	0	1	1	1
	900	—	0	0	0	0	1	1
	1000	0	0	0	0	0	1	1

*Corresponds to 356.2(1).

Definition: *Compact stranding* is the result of a manufacturing process where the standard conductor is compressed to the extent that the interstices (voids between strand wires) are virtually eliminated.

Table C.7 Maximum Number of Conductors or Fixture Wires in Liquidtight Flexible Metal Conduit (LFMC) (*Based on Table 1, Chapter 9*)

Type	Conductor Size (AWG/ kcmil)	CONDUCTORS Metric Designator (Trade Size)									
		16 (½)	21 (¾)	27 (1)	35 (1¼)	41 (1½)	53 (2)	63 (2½)	78 (3)	91 (3½)	103 (4)
RHH, RHW, RHW-2	14	4	7	12	21	27	44	66	102	133	173
	12	3	6	10	17	22	36	55	84	110	144
	10	3	5	8	14	18	29	44	68	89	116
	8	1	2	4	7	9	15	23	36	46	61
	6	1	1	3	6	7	12	18	28	37	48
	4	1	1	2	4	6	9	14	22	29	38
	3	1	1	1	4	5	8	13	19	25	33
	2	1	1	1	3	4	7	11	17	22	29
	1	0	1	1	1	3	5	7	11	14	19
	1/0	0	1	1	1	2	4	6	10	13	16
	2/0	0	1	1	1	1	3	5	8	11	14
	3/0	0	0	1	1	1	3	4	7	9	12
	4/0	0	0	1	1	1	2	4	6	8	10
	250	0	0	0	1	1	1	3	4	6	8
	300	0	0	0	1	1	1	2	4	5	7
	350	0	0	0	1	1	1	2	3	5	6
	400	0	0	0	1	1	1	1	3	4	6
	500	0	0	0	1	1	1	1	3	4	5
	600	0	0	0	0	1	1	1	2	3	4
	700	0	0	0	0	0	1	1	1	3	3
	750	0	0	0	0	0	1	1	1	2	3
	800	0	0	0	0	0	1	1	1	2	3
	900	0	0	0	0	0	1	1	1	2	3
	1000	0	0	0	0	0	1	1	1	1	3
	1250	0	0	0	0	0	0	1	1	1	1
	1500	0	0	0	0	0	0	1	1	1	1
	1750	0	0	0	0	0	0	1	1	1	1
	2000	0	0	0	0	0	0	0	1	1	1
TW	14	9	15	25	44	57	93	140	215	280	365
	12	7	12	19	33	43	71	108	165	215	280
	10	5	9	14	25	32	53	80	123	160	209
	8	3	5	8	14	18	29	44	68	89	116
RHH*, RHW*, RHW-2*, THHW, THW, THW-2	14	6	10	16	29	38	62	93	143	186	243
RHH*, RHW*, RHW-2*, THHW, THW	12	5	8	13	23	30	50	75	115	149	195
	10	3	6	10	18	23	39	58	89	117	152
RHH*, RHW*, RHW-2*, THHW, THW, THW-2	8	1	4	6	11	14	23	35	53	70	91

Table C.7 *Continued*

		CONDUCTORS									
	Conductor Size	Metric Designator (Trade Size)									
Type	(AWG/ kcmil)	16 (½)	21 (¾)	27 (1)	35 (1¼)	41 (1½)	53 (2)	63 (2½)	78 (3)	91 (3½)	103 (4)
RHH*, RHW*, RHW-2*, TW, THW, THHW, THW-2	6	1	3	5	8	11	18	27	41	53	70
	4	1	1	3	6	8	13	20	30	40	52
	3	1	1	3	5	7	11	17	26	34	44
	2	1	1	2	4	6	9	14	22	29	38
	1	1	1	1	3	4	7	10	15	20	26
	1/0	0	1	1	2	3	6	8	13	17	23
	2/0	0	1	1	2	3	5	7	11	15	19
	3/0	0	1	1	1	2	4	6	9	12	16
	4/0	0	0	1	1	1	3	5	8	10	13
	250	0	0	1	1	1	3	4	6	8	11
	300	0	0	1	1	1	2	3	5	7	9
	350	0	0	0	1	1	1	3	5	6	8
	400	0	0	0	1	1	1	3	4	6	7
	500	0	0	0	1	1	1	2	3	5	6
	600	0	0	0	1	1	1	1	3	4	5
	700	0	0	0	0	1	1	1	2	3	4
	750	0	0	0	0	1	1	1	2	3	4
	800	0	0	0	0	1	1	1	2	3	4
	900	0	0	0	0	0	1	1	1	3	3
	1000	0	0	0	0	0	1	1	1	2	3
	1250	0	0	0	0	0	1	1	1	1	2
	1500	0	0	0	0	0	0	1	1	1	2
	1750	0	0	0	0	0	0	1	1	1	1
	2000	0	0	0	0	0	0	1	1	1	1
THHN, THWN, THWN-2	14	13	22	36	63	81	133	201	308	401	523
	12	9	16	26	46	59	97	146	225	292	381
	10	6	10	16	29	37	61	92	141	184	240
	8	3	6	9	16	21	35	53	81	106	138
	6	2	4	7	12	15	25	38	59	76	100
	4	1	2	4	7	9	15	23	36	47	61
	3	1	1	3	6	8	13	20	30	40	52
	2	1	1	3	5	7	11	17	26	33	44
	1	1	1	1	4	5	8	12	19	25	32
	1/0	1	1	1	3	4	7	10	16	21	27
	2/0	0	1	1	2	3	6	8	13	17	23
	3/0	0	1	1	1	3	5	7	11	14	19
	4/0	0	1	1	1	2	4	6	9	12	15
	250	0	0	1	1	1	3	5	7	10	12
	300	0	0	1	1	1	3	4	6	8	11
	350	0	0	1	1	1	2	3	5	7	9
	400	0	0	0	1	1	1	3	5	6	8
	500	0	0	0	1	1	1	2	4	5	7
	600	0	0	0	1	1	1	1	3	4	6
	700	0	0	0	1	1	1	1	3	4	5
	750	0	0	0	0	1	1	1	3	3	5
	800	0	0	0	0	1	1	1	2	3	4
	900	0	0	0	0	1	1	1	2	3	4
	1000	0	0	0	0	0	1	1	1	3	3
FEP, FEPB, PFA, PFAH, TFE	14	12	21	35	61	79	129	195	299	389	507
	12	9	15	25	44	57	94	142	218	284	370
	10	6	11	18	32	41	68	102	156	203	266
	8	3	6	10	18	23	39	58	89	117	152
	6	2	4	7	13	17	27	41	64	83	108
	4	1	3	5	9	12	19	29	44	58	75
	3	1	2	4	7	10	16	24	37	48	63
	2	1	1	3	6	8	13	20	30	40	52

(Continues)

Table C.7 *Continued*

		CONDUCTORS									
	Conductor Size (AWG/ kcmil)	Metric Designator (Trade Size)									
Type		16 (½)	21 (¾)	27 (1)	35 (1¼)	41 (1½)	53 (2)	63 (2½)	78 (3)	91 (3½)	103 (4)
PFA, PFAH, TFE	1	1	1	2	4	5	9	14	21	28	36
PFA, PFAH, TFE, Z	1/0	1	1	1	3	4	7	11	18	23	30
	2/0	1	1	1	3	4	6	9	14	19	25
	3/0	0	1	1	2	3	5	8	12	16	20
	4/0	0	1	1	1	2	4	6	10	13	17
Z	14	20	26	42	73	95	156	235	360	469	611
	12	14	18	30	52	67	111	167	255	332	434
	10	8	11	18	32	41	68	102	156	203	266
	8	5	7	11	20	26	43	64	99	129	168
	6	4	5	8	14	18	30	45	69	90	118
	4	2	3	5	9	12	20	31	48	62	81
	3	2	2	4	7	9	15	23	35	45	59
	2	1	1	3	6	7	12	19	29	38	49
	1	1	1	2	5	6	10	15	23	30	40
XHH, XHHW, XHHW-2, ZW	14	9	15	25	44	57	93	140	215	280	365
	12	7	12	19	33	43	71	108	165	215	280
	10	5	9	14	25	32	53	80	123	160	209
	8	3	5	8	14	18	29	44	68	89	116
	6	1	3	6	10	13	22	33	50	66	86
	4	1	2	4	7	9	16	24	36	48	62
	3	1	1	3	6	8	13	20	31	40	52
	2	1	1	3	5	7	11	17	26	34	44
	1	1	1	1	4	5	8	12	19	25	33
XHH, XHHW, XHHW-2	1/0	1	1	1	3	4	7	10	16	21	28
	2/0	0	1	1	2	3	6	9	13	17	23
	3/0	0	1	1	1	3	5	7	11	14	19
	4/0	0	1	1	1	2	4	6	9	12	16
	250	0	0	1	1	1	3	5	7	10	13
	300	0	0	1	1	1	3	4	6	8	11
	350	0	0	1	1	1	2	3	5	7	10
	400	0	0	0	1	1	1	3	5	6	8
	500	0	0	0	1	1	1	2	4	5	7
	600	0	0	0	1	1	1	1	3	4	6
	700	0	0	0	1	1	1	1	3	4	5
	750	0	0	0	0	1	1	1	3	3	5
	800	0	0	0	0	1	1	1	2	3	4
	900	0	0	0	0	1	1	1	2	3	4
	1000	0	0	0	0	0	1	1	1	3	3
	1250	0	0	0	0	0	1	1	1	1	3
	1500	0	0	0	0	0	1	1	1	1	2
	1750	0	0	0	0	0	0	1	1	1	2
	2000	0	0	0	0	0	0	1	1	1	2

Table C.7 *Continued*

	Conductor Size	**FIXTURE WIRES**					
		Metric Designator (Trade Size)					
Type	**(AWG/ kcmil)**	**16** **(½)**	**21** **(¾)**	**27** **(1)**	**35** **(1¼)**	**41** **(1½)**	**53** **(2)**
FFH-2, RFH-2, RFHH-3	18	8	15	24	42	54	89
	16	7	12	20	35	46	75
SF-2, SFF-2	18	11	19	30	53	69	113
	16	9	15	25	44	57	93
	14	7	12	20	35	46	75
SF-1, SFF-1	18	19	33	53	94	122	199
RFH-1, RFHH-2, TF, TFF, XF, XFF	18	14	24	39	69	90	147
RFHH-2, TF, TFF, XF, XFF	16	11	20	32	56	72	119
XF, XFF	14	9	15	25	44	57	93
TFN, TFFN	18	23	39	63	111	144	236
	16	17	30	48	85	110	180
PF, PFF, PGF, PGFF, PAF, PTF, PTFF, PAFF	18	21	37	60	105	136	223
	16	16	29	46	81	105	173
	14	12	21	35	61	79	129
HF, HFF, ZF, ZFF, ZHF	18	28	48	77	136	176	288
	16	20	35	57	100	129	212
	14	15	26	42	73	95	156
KF-2, KFF-2	18	40	70	112	197	255	418
	16	28	49	79	139	180	295
	14	19	34	54	95	123	202
	12	13	23	37	65	85	139
	10	9	15	25	44	57	93
KF-1, KFF-1	18	48	83	134	235	304	499
	16	34	58	94	165	214	350
	14	23	39	63	111	144	236
	12	15	26	42	73	95	156
	10	10	17	27	48	62	102
XF, XFF	12	5	8	13	23	30	50
	10	3	6	10	18	23	39

Notes:

1. This table is for concentric stranded conductors only. For compact stranded conductors, Table C.7(A) should be used.

2. Two-hour fire-rated RHH cable has ceramifiable insulation which has much larger diameters than other RHH wires. Consult manufacturer's conduit fill tables.

*Types RHH, RHW, and RHW-2 without outer covering.

Table C.7(A) Maximum Number of Compact Conductors in Liquidtight Flexible Metal Conduit (LFMC) *(Based on Table 1, Chapter 9)*

	Conductor Size (AWG/ kcmil)	COMPACT CONDUCTORS										
		Metric Designator (Trade Size)										
Type		12 (⅜)	16 (½)	21 (¾)	27 (1)	35 (1¼)	41 (1½)	53 (2)	63 (2½)	78 (3)	91 (3½)	103 (4)
THW,	8	1	2	4	7	12	15	25	38	58	76	99
THW-2,	6	1	1	3	5	9	12	19	29	45	59	77
THHW	4	1	1	2	4	7	9	14	22	34	44	57
	2	1	1	1	3	5	6	11	16	25	32	42
	1	0	1	1	1	3	4	7	11	17	23	30
	1/0	0	1	1	1	3	4	6	10	15	20	26
	2/0	0	0	1	1	2	3	5	8	13	16	21
	3/0	0	0	1	1	1	3	4	7	11	14	18
	4/0	0	0	1	1	1	2	4	6	9	12	15
	250	0	0	0	1	1	1	3	4	7	9	12
	300	0	0	0	1	1	1	2	4	6	8	10
	350	0	0	0	1	1	1	2	3	5	7	9
	400	0	0	0	0	1	1	1	3	5	6	8
	500	0	0	0	0	1	1	1	3	4	5	7
	600	0	0	0	0	1	1	1	1	3	4	6
	700	0	0	0	0	1	1	1	1	3	4	5
	750	0	0	0	0	0	1	1	1	3	3	5
	900	—	0	0	0	0	1	1	1	2	3	4
	1000	0	0	0	0	0	1	1	1	1	3	4
THHN,	8	—	—	—	—	—	—	—	—	—	—	—
THWN,	6	1	2	4	7	13	17	28	43	66	86	112
THWN-2	4	1	1	3	4	8	11	17	26	41	53	69
	2	1	1	1	3	6	7	12	19	29	38	50
	1	0	1	1	2	4	6	9	14	22	28	37
	1/0	0	1	1	1	4	5	8	12	19	24	32
	2/0	0	1	1	1	3	4	6	10	15	20	26
	3/0	0	0	1	1	2	3	5	8	13	17	22
	4/0	0	0	1	1	1	3	4	7	10	14	18
	250	0	0	1	1	1	1	3	5	8	11	14
	300	0	0	0	1	1	1	3	4	7	9	12
	350	0	0	0	1	1	1	2	4	6	8	11
	400	0	0	0	1	1	1	2	3	5	7	9
	500	0	0	0	0	1	1	1	3	5	6	8
	600	0	0	0	0	1	1	1	2	4	5	6
	700	0	0	0	0	1	1	1	1	3	4	6
	750	0	0	0	0	1	1	1	1	3	4	5
	900	—	0	0	0	0	1	1	1	2	3	4
	1000	0	0	0	0	0	1	1	1	2	3	4
XHHW,	8	1	3	5	9	15	20	33	49	76	98	129
XHHW-2	6	1	2	4	6	11	15	24	37	56	73	95
	4	1	1	3	4	8	11	17	26	41	53	69
	2	1	1	1	3	6	7	12	19	29	38	50
	1	0	1	1	2	4	6	9	14	22	28	37
	1/0	0	1	1	1	4	5	8	12	19	24	32
	2/0	0	1	1	1	3	4	7	10	16	20	27
	3/0	0	0	1	1	2	3	5	8	13	17	22
	4/0	0	0	1	1	1	3	4	7	11	14	18
	250	0	0	1	1	1	1	3	5	8	11	15
	300	0	0	0	1	1	1	3	5	7	9	12
	350	0	0	0	1	1	1	3	4	6	8	11
	400	0	0	0	1	1	1	2	4	6	7	10
	500	0	0	0	0	1	1	1	3	5	6	8
	600	0	0	0	0	1	1	1	2	4	5	6
	700	0	0	0	0	1	1	1	1	3	4	6
	750	0	0	0	0	1	1	1	1	3	4	5
	900	—	0	0	0	0	1	1	2	2	3	4
	1000	0	0	0	0	0	1	1	1	2	3	4

Definition: *Compact stranding* is the result of a manufacturing process where the standard conductor is compressed to the extent that the interstices (voids between strand wires) are virtually eliminated.

Table C.8 Maximum Number of Conductors or Fixture Wires in Rigid Metal Conduit (RMC)
(*Based on Table 1, Chapter 9*)

		CONDUCTORS											
	Conductor Size	Metric Designator (Trade Size)											
Type	(AWG/ kcmil)	16 (½)	21 (¾)	27 (1)	35 (1¼)	41 (1½)	53 (2)	63 (2½)	78 (3)	91 (3½)	103 (4)	129 (5)	155 (6)
RHH,	14	4	7	12	21	28	46	66	102	136	176	276	398
RHW,	12	3	6	10	17	23	38	55	85	113	146	229	330
RHW-2	10	3	5	8	14	19	31	44	68	91	118	185	267
	8	1	2	4	7	10	16	23	36	48	61	97	139
	6	1	1	3	6	8	13	18	29	38	49	77	112
	4	1	1	2	4	6	10	14	22	30	38	60	87
	3	1	1	2	4	5	9	12	19	26	34	53	76
	2	1	1	1	3	4	7	11	17	23	29	46	66
	1	0	1	1	1	3	5	7	11	15	19	30	44
	1/0	0	1	1	1	2	4	6	10	13	17	26	38
	2/0	0	1	1	1	2	4	5	8	11	14	23	33
	3/0	0	0	1	1	1	3	4	7	10	12	20	28
	4/0	0	0	1	1	1	3	4	6	8	11	17	24
	250	0	0	0	1	1	1	3	4	6	8	13	18
	300	0	0	0	1	1	1	2	4	5	7	11	16
	350	0	0	0	1	1	1	2	4	5	6	10	15
	400	0	0	0	1	1	1	1	3	4	6	9	13
	500	0	0	0	1	1	1	1	3	4	5	8	11
	600	0	0	0	0	1	1	1	2	3	4	6	9
	700	0	0	0	0	1	1	1	1	3	4	6	8
	750	0	0	0	0	0	1	1	1	3	3	5	8
	800	0	0	0	0	0	1	1	1	2	3	5	7
	900	0	0	0	0	0	1	1	1	2	3	5	7
	1000	0	0	0	0	0	1	1	1	1	3	4	6
	1250	0	0	0	0	0	0	1	1	1	1	3	5
	1500	0	0	0	0	0	0	1	1	1	1	3	4
	1750	0	0	0	0	0	0	1	1	1	1	2	4
	2000	0	0	0	0	0	0	0	1	1	1	2	3
TW	14	9	15	25	44	59	98	140	216	288	370	581	839
	12	7	12	19	33	45	75	107	165	221	284	446	644
	10	5	9	14	25	34	56	80	123	164	212	332	480
	8	3	5	8	14	19	31	44	68	91	118	185	267
RHH*, RHW*, RHW-2* THHW, THW, THW-2	14	6	10	17	29	39	65	93	143	191	246	387	558
RHH*, RHW*, RHW-2*, THHW, THW	12	5	8	13	23	32	52	75	115	154	198	311	448
	10	3	6	10	18	25	41	58	90	120	154	242	350
RHH*, RHW*, RHW-2*, THHW, THW, THW-2	8	1	4	6	11	15	24	35	54	72	92	145	209

(Continues)

Table C.8 *Continued*

		CONDUCTORS											
	Conductor Size	Metric Designator (Trade Size)											
Type	(AWG/ kcmil)	16 (½)	21 (¾)	27 (1)	35 (1¼)	41 (1½)	53 (2)	63 (2½)	78 (3)	91 (3½)	103 (4)	129 (5)	155 (6)
RHH*,	6	1	3	5	8	11	18	27	41	55	71	111	160
RHW*,	4	1	1	3	6	8	14	20	31	41	53	83	120
RHW-2*,	3	1	1	3	5	7	12	17	26	35	45	71	103
TW,	2	1	1	2	4	6	10	14	22	30	38	60	87
THW,	1	1	1	1	3	4	7	10	15	21	27	42	61
THHW,	1/0	0	1	1	2	3	6	8	13	18	23	36	52
THW-2	2/0	0	1	1	2	3	5	7	11	15	19	31	44
	3/0	0	1	1	1	2	4	6	9	13	16	26	37
	4/0	0	0	1	1	1	3	5	8	10	14	21	31
	250	0	0	1	1	1	3	4	6	8	11	17	25
	300	0	0	1	1	1	2	3	5	7	9	15	22
	350	0	0	0	1	1	1	3	5	6	8	13	19
	400	0	0	0	1	1	1	3	4	6	7	12	17
	500	0	0	0	1	1	1	2	3	5	6	10	14
	600	0	0	0	1	1	1	1	3	4	5	8	12
	700	0	0	0	0	1	1	1	2	3	4	7	10
	750	0	0	0	0	1	1	1	2	3	4	7	10
	800	0	0	0	0	1	1	1	2	3	4	6	9
	900	0	0	0	0	1	1	1	1	3	4	6	8
	1000	0	0	0	0	0	1	1	1	2	3	5	8
	1250	0	0	0	0	0	1	1	1	1	2	4	6
	1500	0	0	0	0	0	1	1	1	1	2	3	5
	1750	0	0	0	0	0	0	1	1	1	1	3	4
	2000	0	0	0	0	0	0	1	1	1	1	3	4
THHN,	14	13	22	36	63	85	140	200	309	412	531	833	1202
THWN,	12	9	16	26	46	62	102	146	225	301	387	608	877
THWN-2	10	6	10	17	29	39	64	92	142	189	244	383	552
	8	3	6	9	16	22	37	53	82	109	140	221	318
	6	2	4	7	12	16	27	38	59	79	101	159	230
	4	1	2	4	7	10	16	23	36	48	62	98	141
	3	1	1	3	6	8	14	20	31	41	53	83	120
	2	1	1	3	5	7	11	17	26	34	44	70	100
	1	1	1	1	4	5	8	12	19	25	33	51	74
	1/0	1	1	1	3	4	7	10	16	21	27	43	63
	2/0	0	1	1	2	3	6	8	13	18	23	36	52
	3/0	0	1	1	1	3	5	7	11	15	19	30	43
	4/0	0	1	1	1	2	4	6	9	12	16	25	36
	250	0	0	1	1	1	3	5	7	10	13	20	29
	300	0	0	1	1	1	3	4	6	8	11	17	25
	350	0	0	1	1	1	2	3	5	7	10	15	22
	400	0	0	1	1	1	2	3	5	7	8	13	20
	500	0	0	0	1	1	1	2	4	5	7	11	16
	600	0	0	0	1	1	1	1	3	4	6	9	13
	700	0	0	0	1	1	1	1	3	4	5	8	11
	750	0	0	0	0	1	1	1	3	4	5	7	11
	800	0	0	0	0	1	1	1	2	3	4	7	10
	900	0	0	0	0	1	1	1	2	3	4	6	9
	1000	0	0	0	0	1	1	1	1	3	4	6	8
FEP,	14	12	22	35	61	83	136	194	300	400	515	808	1166
FEPB,	12	9	16	26	44	60	99	142	219	292	376	590	851
PFA,	10	6	11	18	32	43	71	102	157	209	269	423	610
PFAH,	8	3	6	10	18	25	41	58	90	120	154	242	350
TFE	6	2	4	7	13	17	29	41	64	85	110	172	249
	4	1	3	5	9	12	20	29	44	59	77	120	174
	3	1	2	4	7	10	17	24	37	50	64	100	145
	2	1	1	3	6	8	14	20	31	41	53	83	120

Table C.8 *Continued*

						CONDUCTORS						
	Conductor Size				Metric Designator (Trade Size)							
Type	(AWG/ kcmil)	16 (½)	21 (¾)	27 (1)	35 (1¼)	41 (1½)	53 (2)	63 (2½)	78 (3)	91 (3½)	103 (4)	129 (5)	155 (6)
PFA, PFAH, TFE	1	1	1	2	4	6	9	14	21	28	37	57	83
PFA, PFAH, TFE, Z	1/0	1	1	1	3	5	8	11	18	24	30	48	69
	2/0	1	1	1	3	4	6	9	14	19	25	40	57
	3/0	0	1	1	2	3	5	8	12	16	21	33	47
	4/0	0	1	1	1	2	4	6	10	13	17	27	39
Z	14	15	26	42	73	100	164	234	361	482	621	974	1405
	12	10	18	30	52	71	116	166	256	342	440	691	997
	10	6	11	18	32	43	71	102	157	209	269	423	610
	8	4	7	11	20	27	45	64	99	132	170	267	386
	6	3	5	8	14	19	31	45	69	93	120	188	271
	4	1	3	5	9	13	22	31	48	64	82	129	186
	3	1	2	4	7	9	16	22	35	47	60	94	136
	2	1	1	3	6	8	13	19	29	39	50	78	113
	1	1	1	2	5	6	10	15	23	31	40	63	92
XHH, XHHW, XHHW-2 ZW	14	9	15	25	44	59	98	140	216	288	370	581	839
	12	7	12	19	33	45	75	107	165	221	284	446	644
	10	5	9	14	25	34	56	80	123	164	212	332	480
	8	3	5	8	14	19	31	44	68	91	118	185	267
	6	1	3	6	10	14	23	33	51	68	87	137	197
	4	1	2	4	7	10	16	24	37	49	63	99	143
	3	1	1	3	6	8	14	20	31	41	53	84	121
	2	1	1	3	5	7	12	17	26	35	45	70	101
	1	1	1	1	4	5	9	12	19	26	33	52	76
XHH, XHHW, XHHW-2	1/0	1	1	1	3	4	7	10	16	22	28	44	64
	2/0	0	1	1	2	3	6	9	13	18	23	37	53
	3/0	0	1	1	1	3	5	7	11	15	19	30	44
	4/0	0	1	1	1	2	4	6	9	12	16	25	36
	250	0	0	1	1	1	3	5	7	10	13	20	30
	300	0	0	1	1	1	3	4	6	9	11	18	25
	350	0	0	1	1	1	2	3	6	7	10	15	22
	400	0	0	1	1	1	2	3	5	7	9	14	20
	500	0	0	0	1	1	1	2	4	5	7	11	16
	600	0	0	0	1	1	1	1	3	4	6	9	13
	700	0	0	0	1	1	1	1	3	4	5	8	11
	750	0	0	0	0	1	1	1	3	4	5	7	11
	800	0	0	0	0	1	1	1	2	3	4	7	10
	900	0	0	0	0	1	1	1	2	3	4	6	9
	1000	0	0	0	0	1	1	1	1	3	4	6	8
	1250	0	0	0	0	0	1	1	1	2	3	4	6
	1500	0	0	0	0	0	1	1	1	1	2	4	5
	1750	0	0	0	0	0	0	1	1	1	1	3	5
	2000	0	0	0	0	0	0	1	1	1	1	3	4

(Continues)

Table C.8 *Continued*

	Conductor Size (AWG/ kcmil)	Metric Designator (Trade Size)					
Type		**16** (½)	**21** (¾)	**27** (1)	**35** (1¼)	**41** (1½)	**53** (2)
						FIXTURE WIRES	

Type	Conductor Size (AWG/ kcmil)	**16** (½)	**21** (¾)	**27** (1)	**35** (1¼)	**41** (1½)	**53** (2)
FFH-2,	18	8	15	24	42	57	94
RFH-2, RFHH-3	16	7	12	20	35	48	79
SF-2, SFF-2	18	11	19	31	53	72	118
	16	9	15	25	44	59	98
	14	7	12	20	35	48	79
SF-1, SFF-1	18	19	33	54	94	127	209
RFH-1, RFHH-2, TF, TFF, XF, XFF	18	14	25	40	69	94	155
RFHH-2, TF, TFF, XF, XFF	16	11	20	32	56	76	125
XF, XFF	14	9	15	25	44	59	98
TFN, TFFN	18	23	40	64	111	150	248
	16	17	30	49	84	115	189
PF, PFF, PGF, PGFF, PAF, PTF, PTFF, PAFF	18	21	38	61	105	143	235
	16	16	29	47	81	110	181
	14	12	22	35	61	83	136
HF, HFF, ZF, ZFF, ZHF	18	28	48	79	135	184	303
	16	20	36	58	100	136	223
	14	15	26	42	73	100	164
KF-2, KFF-2	18	40	71	114	197	267	439
	16	28	50	80	138	188	310
	14	19	34	55	95	129	213
	12	13	23	38	65	89	146
	10	9	15	25	44	59	98
KF-1, KFF-1	18	48	84	136	235	318	524
	16	34	59	96	165	224	368
	14	23	40	64	111	150	248
	12	15	26	42	73	100	164
	10	10	17	28	48	65	107
XF, XFF	12	5	8	13	23	32	52
	10	3	6	10	18	25	41

Notes:

1. This table is for concentric stranded conductors only. For compact stranded conductors, Table C.8(A) should be used.

2. Two-hour fire-rated RHH cable has ceramifiable insulation which has much larger diameters than other RHH wires. Consult manufacturer's conduit fill tables.

*Types RHH, RHW, and RHW-2 without outer covering.

Table C.8(A) Maximum Number of Compact Conductors in Rigid Metal Conduit (RMC)
(*Based on Table 1, Chapter 9*)

	Conductor Size	COMPACT CONDUCTORS											
		Metric Designator (Trade Size)											
Type	(AWG/ kcmil)	16 (½)	21 (¾)	27 (1)	35 (1¼)	41 (1½)	53 (2)	63 (2½)	78 (3)	91 (3½)	103 (4)	129 (5)	155 (6)
THW,	8	2	4	7	12	16	26	38	59	78	101	158	228
THW-2,	6	1	3	5	9	12	20	29	45	60	78	122	176
THHW	4	1	2	4	7	9	15	22	34	45	58	91	132
	2	1	1	3	5	7	11	16	25	33	43	67	97
	1	1	1	1	3	5	8	11	17	23	30	47	68
	1/0	1	1	1	3	4	7	10	15	20	26	41	59
	2/0	0	1	1	2	3	6	8	13	17	22	34	50
	3/0	0	1	1	1	3	5	7	11	14	19	29	42
	4/0	0	1	1	1	2	4	6	9	12	15	24	35
	250	0	0	1	1	1	3	4	7	9	12	19	28
	300	0	0	1	1	1	3	4	6	8	11	17	24
	350	0	0	1	1	1	2	3	5	7	9	15	22
	400	0	0	1	1	1	1	3	5	7	8	13	20
	500	0	0	0	1	1	1	3	4	5	7	11	17
	600	0	0	0	1	1	1	1	3	4	6	9	13
	700	0	0	0	1	1	1	1	3	4	5	8	12
	750	0	0	0	0	1	1	1	3	4	5	7	11
	900	0	0	0	0	1	1	1	2	3	4	6	9
	1000	0	0	0	0	1	1	1	1	3	4	6	9
THHN,	8	—	—	—	—	—	—	—	—	—	—	—	—
THWN,	6	2	5	8	13	18	30	43	66	88	114	179	258
THWN-2	4	1	3	5	8	11	18	26	41	55	70	110	159
	2	1	1	3	6	8	13	19	29	39	50	79	114
	1	1	1	2	4	6	10	14	22	29	38	60	86
	1/0	1	1	1	4	5	8	12	19	25	32	51	73
	2/0	1	1	1	3	4	7	10	15	21	26	42	60
	3/0	0	1	1	2	3	6	8	13	17	22	35	51
	4/0	0	1	1	1	3	5	7	10	14	18	29	42
	250	0	1	1	1	2	4	5	8	11	14	23	33
	300	0	0	1	1	1	3	4	7	10	12	20	28
	350	0	0	1	1	1	3	4	6	8	11	17	25
	400	0	0	1	1	1	2	3	5	7	10	15	22
	500	0	0	0	1	1	1	3	5	6	8	13	19
	600	0	0	0	1	1	1	2	4	5	6	10	15
	700	0	0	0	1	1	1	1	3	4	6	9	13
	750	0	0	0	1	1	1	1	3	4	5	9	13
	900	0	0	0	0	1	1	1	2	3	4	6	9
	1000	0	0	0	0	1	1	1	2	3	4	6	9
XHHW,	8	3	5	9	15	21	34	49	76	101	130	205	296
XHHW-2	6	2	4	6	11	15	25	36	56	75	97	152	220
	4	1	3	5	8	11	18	26	41	55	70	110	159
	2	1	1	3	6	8	13	19	29	39	50	79	114
	1	1	1	2	4	6	10	14	22	29	38	60	86
	1/0	1	1	1	4	5	8	12	19	25	32	51	73
	2/0	1	1	1	3	4	7	10	16	21	27	43	62
	3/0	0	1	1	2	3	6	8	13	17	22	35	51
	4/0	0	1	1	1	3	5	7	11	14	19	29	42
	250	0	1	1	1	2	4	5	8	11	15	23	34
	300	0	0	1	1	1	3	5	7	10	13	20	29
	350	0	0	1	1	1	3	4	6	9	11	18	25
	400	0	0	1	1	1	2	4	6	8	10	16	23
	500	0	0	0	1	1	1	3	5	6	8	13	19
	600	0	0	0	1	1	1	2	4	5	7	10	15
	700	0	0	0	1	1	1	1	3	4	6	9	13
	750	0	0	0	1	1	1	1	3	4	5	8	12
	900	0	0	0	0	1	1	2	2	3	5	7	10
	1000	0	0	0	0	1	1	1	2	3	4	7	10

Definition: *Compact stranding* is the result of a manufacturing process where the standard conductor is compressed to the extent that the interstices (voids between strand wires) are virtually eliminated.

Table C.9 Maximum Number of Conductors or Fixture Wires in Rigid PVC Conduit, Schedule 80 (*Based on Table 1, Chapter 9*)

		CONDUCTORS											
	Conductor Size	Metric Designator (Trade Size)											
Type	(AWG/ kcmil)	16 (½)	21 (¾)	27 (1)	35 (1¼)	41 (1½)	53 (2)	63 (2½)	78 (3)	91 (3½)	103 (4)	129 (5)	155 (6)
RHH, RHW, RHW-2	14	3	5	9	17	23	39	56	88	118	153	243	349
	12	2	4	7	14	19	32	46	73	98	127	202	290
	10	1	3	6	11	15	26	37	59	79	103	163	234
	8	1	1	3	6	8	13	19	31	41	54	85	122
	6	1	1	2	4	6	11	16	24	33	43	68	98
	4	1	1	1	3	5	8	12	19	26	33	53	77
	3	0	1	1	3	4	7	11	17	23	29	47	67
	2	0	1	1	3	4	6	9	14	20	25	41	58
	1	0	1	1	1	2	4	6	9	13	17	27	38
	1/0	0	0	1	1	1	3	5	8	11	15	23	33
	2/0	0	0	1	1	1	3	4	7	10	13	20	29
	3/0	0	0	1	1	1	3	4	6	8	11	17	25
	4/0	0	0	0	1	1	2	3	5	7	9	15	21
	250	0	0	0	1	1	1	2	4	5	7	11	16
	300	0	0	0	1	1	1	2	3	5	6	10	14
	350	0	0	0	1	1	1	1	3	4	5	9	13
	400	0	0	0	0	1	1	1	3	4	5	8	12
	500	0	0	0	0	1	1	1	2	3	4	7	10
	600	0	0	0	0	0	1	1	1	3	3	6	8
	700	0	0	0	0	0	1	1	1	2	3	5	7
	750	0	0	0	0	0	1	1	1	2	3	5	7
	800	0	0	0	0	0	1	1	1	2	3	4	7
	1000	0	0	0	0	0	1	1	1	1	2	4	5
	1250	0	0	0	0	0	0	1	1	1	1	3	4
	1500	0	0	0	0	0	0	1	1	1	1	2	4
	1750	0	0	0	0	0	0	0	1	1	1	2	3
	2000	0	0	0	0	0	0	0	1	1	1	1	3
TW	14	6	11	20	35	49	82	118	185	250	324	514	736
	12	5	9	15	27	38	63	91	142	192	248	394	565
	10	3	6	11	20	28	47	67	106	143	185	294	421
	8	1	3	6	11	15	26	37	59	79	103	163	234
RHH*, RHW*, RHW-2*, THHW, THW, THW-2	14	4	8	13	23	32	55	79	123	166	215	341	490
RHH*, RHW*, RHW-2*, THHW, THW	12	3	6	10	19	26	44	63	99	133	173	274	394
	10	2	5	8	15	20	34	49	77	104	135	214	307
RHH*, RHW*, RHW-2*, THHW, THW, THW-2	8	1	3	5	9	12	20	29	46	62	81	128	184

Table C.9 *Continued*

		CONDUCTORS											
	Conductor Size	Metric Designator (Trade Size)											
Type	(AWG/ kcmil)	16 (½)	21 (¾)	27 (1)	35 (1¼)	41 (1½)	53 (2)	63 (2½)	78 (3)	91 (3½)	103 (4)	129 (5)	155 (6)
RHH*,	6	1	1	3	7	9	16	22	35	48	62	98	141
RHW*,	4	1	1	3	5	7	12	17	26	35	46	73	105
RHW-2*,	3	1	1	2	4	6	10	14	22	30	39	63	90
TW,	2	1	1	1	3	5	8	12	19	26	33	53	77
THW,	1	0	1	1	2	3	6	8	13	18	23	37	54
THHW,	1/0	0	1	1	1	3	5	7	11	15	20	32	46
THW-2	2/0	0	1	1	1	2	4	6	10	13	17	27	39
	3/0	0	0	1	1	1	3	5	8	11	14	23	33
	4/0	0	0	1	1	1	3	4	7	9	12	19	27
	250	0	0	0	1	1	2	3	5	7	9	15	22
	300	0	0	0	1	1	1	3	5	6	8	13	19
	350	0	0	0	1	1	1	2	4	6	7	12	17
	400	0	0	0	1	1	1	2	4	5	7	10	15
	500	0	0	0	1	1	1	1	3	4	5	9	13
	600	0	0	0	0	1	1	1	2	3	4	7	10
	700	0	0	0	0	1	1	1	2	3	4	6	9
	750	0	0	0	0	0	1	1	1	3	4	6	8
	800	0	0	0	0	0	1	1	1	3	3	6	8
	900	0	0	0	0	0	1	1	1	2	3	5	7
	1000	0	0	0	0	0	1	1	1	2	3	5	7
	1250	0	0	0	0	0	1	1	1	1	2	4	5
	1500	0	0	0	0	0	0	1	1	1	1	3	4
	1750	0	0	0	0	0	0	1	1	1	1·	3	4
	2000	0	0	0	0	0	0	0	1	1	1	2	3
THHN,	14	9	17	28	51	70	118	170	265	358	464	736	1055
THWN,	12	6	12	20	37	51	86	124	193	261	338	537	770
THWN-2	10	4	7	13	23	32	54	78	122	164	213	338	485
	8	2	4	7	13	18	31	45	70	95	123	195	279
	6	1	3	5	9	13	22	32	51	68	89	141	202
	4	1	1	3	6	8	14	20	31	42	54	86	124
	3	1	1	3	5	7	12	17	26	35	46	73	105
	2	1	1	2	4	6	10	14	22	30	39	61	88
	1	0	1	1	3	4	7	10	16	22	29	45	65
	1/0	0	1	1	2	3	6	9	14	18	24	38	55
	2/0	0	1	1	1	3	5	7	11	15	20	32	46
	3/0	0	1	1	1	2	4	6	9	13	17	26	38
	4/0	0	0	1	1	1	3	5	8	10	14	22	31
	250	0	0	1	1	1	3	4	6	8	11	18	25
	300	0	0	0	1	1	2	3	5	7	9	15	22
	350	0	0	0	1	1	1	3	5	6	8	13	19
	400	0	0	0	1	1	1	3	4	6	7	12	17
	500	0	0	0	1	1	1	2	3	5	6	10	14
	600	0	0	0	0	1	1	1	3	4	5	8	12
	700	0	0	0	0	1	1	1	2	3	4	7	10
	750	0	0	0	0	1	1	1	2	3	4	7	9
	800	0	0	0	0	1	1	1	2	3	4	6	9
	900	0	0	0	0	0	1	1	1	3	3	6	8
	1000	0	0	0	0	0	1	1	1	2	3	5	7
FEP,	14	8	16	27	49	68	115	164	257	347	450	714	1024
FEPB,	12	6	12	20	36	50	84	120	188	253	328	521	747
PFA,	10	4	8	14	26	36	60	86	135	182	235	374	536
PFAH,	8	2	5	8	15	20	34	49	77	104	135	214	307
TFE	6	1	3	6	10	14	24	35	55	74	96	152	218
	4	1	2	4	7	10	17	24	38	52	67	106	153
	3	1	1	3	6	8	14	20	32	43	56	89	127
	2	1	1	3	5	7	12	17	26	35	46	73	105

(Continues)

Table C.9 *Continued*

		CONDUCTORS										
	Conductor Size (AWG/ kcmil)	Metric Designator (Trade Size)										
Type		16 (½)	21 (¾)	27 (1)	35 (1¼)	41 (1½)	53 (2)	63 (2½)	78 (3)	91 (3½)	103 (4)	129 (5)	155 (6)
PFA, PFAH, TFE	1	1	1	1	3	5	8	11	18	25	32	51	73
PFA, PFAH, TFE, Z	1/0	0	1	1	3	4	7	10	15	20	27	42	61
	2/0	0	1	1	2	3	5	8	12	17	22	35	50
	3/0	0	1	1	1	2	4	6	10	14	18	29	41
	4/0	0	0	1	1	1	4	5	8	11	15	24	34
Z	14	10	19	33	59	82	138	198	310	418	542	860	1233
	12	7	14	23	42	58	98	141	220	297	385	610	875
	10	4	8	14	26	36	60	86	135	182	235	374	536
	8	3	5	9	16	22	38	54	85	115	149	236	339
	6	2	4	6	11	16	26	38	60	81	104	166	238
	4	1	2	4	8	11	18	26	41	55	72	114	164
	3	1	2	3	5	8	13	19	30	40	52	83	119
	2	1	1	2	5	6	11	16	25	33	43	69	99
	1	0	1	2	4	5	9	13	20	27	35	56	80
XHH, XHHW, XHHW-2, ZW	14	6	11	20	35	49	82	118	185	250	324	514	736
	12	5	9	15	27	38	63	91	142	192	248	394	565
	10	3	6	11	20	28	47	67	106	143	185	294	421
	8	1	3	6	11	15	26	37	59	79	103	163	234
	6	1	2	4	8	11	19	28	43	59	76	121	173
	4	1	1	3	6	8	14	20	31	42	55	87	125
	3	1	1	3	5	7	12	17	26	36	47	74	106
	2	1	1	2	4	6	10	14	22	30	39	62	89
XHH, XHHW, XHHW-2	1	0	1	1	3	4	7	10	16	22	29	46	66
	1/0	0	1	1	2	3	6	9	14	19	24	39	56
	2/0	0	1	1	1	3	5	7	11	16	20	32	46
	3/0	0	1	1	1	2	4	6	9	13	17	27	38
	4/0	0	0	1	1	1	3	5	8	11	14	22	32
	250	0	0	1	1	1	3	4	6	9	11	18	26
	300	0	0	1	1	1	2	3	5	7	10	15	22
	350	0	0	0	1	1	1	3	5	6	8	14	20
	400	0	0	0	1	1	1	3	4	6	7	12	17
	500	0	0	0	1	1	1	2	3	5	6	10	14
	600	0	0	0	0	1	1	1	3	4	5	8	11
	700	0	0	0	0	1	1	1	2	3	4	7	10
	750	0	0	0	0	1	1	1	2	3	4	6	9
	800	0	0	0	0	1	1	1	1	3	4	6	9
	900	0	0	0	0	0	1	1	—	3	3	5	8
	1000	0	0	0	0	0	1	1	1	2	3	5	7
	1250	0	0	0	0	0	1	1	1	1	2	4	6
	1500	0	0	0	0	0	0	1	1	1	1	3	5
	1750	0	0	0	0	0	0	1	1	1	1	3	4
	2000	0	0	0	0	0	0	1	1	1	1	2	4

Table C.9 *Continued*

		FIXTURE WIRES					
	Conductor Size	Metric Designator (Trade Size)					
Type	(AWG/ kcmil)	16 (½)	21 (¾)	27 (1)	35 (1¼)	41 (1½)	53 (2)
FFH-2, RFH-2, RFHH-3	18	6	11	19	34	47	79
	16	5	9	16	28	39	67
SF-2, SFF-2	18	7	14	24	43	59	100
	16	6	11	20	35	49	82
	14	5	9	16	28	39	67
SF-1, SFF-1	18	13	25	42	76	105	177
RFH-1, RFHH-2, TF, TFF, XF, XFF	18	10	18	31	56	77	130
RFHH-2, TF, TFF, XF, XFF	16	8	15	25	45	62	105
XF, XFF	14	6	11	20	35	49	82
TFN, TFFN	18	16	29	50	90	124	209
	16	12	22	38	68	95	159
PF, PFF, PGF, PGFF, PAF, PTF, PTFF, PAFF	18	15	28	47	85	118	198
	16	11	22	36	66	91	153
	14	8	16	27	49	68	115
HF, HFF, ZF, ZFF, ZHF	18	19	36	61	110	152	255
	16	14	27	45	81	112	188
	14	10	19	33	59	82	138
KF-2, KFF-2	18	28	53	88	159	220	371
	16	19	37	62	112	155	261
	14	13	25	43	77	107	179
	12	9	17	29	53	73	123
	10	6	11	20	35	49	82
KF-1, KFF-1	18	33	63	106	190	263	442
	16	23	44	74	133	185	310
	14	16	29	50	90	124	209
	12	10	19	33	59	82	138
	10	7	13	21	39	54	90
XF, XFF	12	3	6	10	19	26	44
	10	2	5	8	15	20	34

Notes:

1. This table is for concentric stranded conductors only. For compact stranded conductors, Table C.9(A) should be used.

2. Two-hour fire-rated RHH cable has ceramifiable insulation which has much larger diameters than other RHH wires. Consult manufacturer's conduit fill tables.

*Types RHH, RHW, and RHW-2 without outer covering.

Table C.9(A) Maximum Number of Compact Conductors in Rigid PVC Conduit, Schedule 80
(*Based on Table 1, Chapter 9*)

	Conductor Size (AWG/ kcmil)	COMPACT CONDUCTORS											
		Metric Designator (Trade Size)											
Type		16 (½)	21 (¾)	27 (1)	35 (1¼)	41 (1½)	53 (2)	63 (2½)	78 (3)	91 (3½)	103 (4)	129 (5)	155 (6)
THW, THW-2, THHW	8	1	3	5	9	13	22	32	50	68	88	140	200
	6	1	2	4	7	10	17	25	39	52	68	108	155
	4	1	1	3	5	7	13	18	29	39	51	81	116
	2	1	1	1	4	5	9	13	21	29	37	60	85
	1	0	1	1	3	4	6	9	15	20	26	42	60
	1/0	0	1	1	2	3	6	8	13	17	23	36	52
	2/0	0	1	1	1	3	5	7	11	15	19	30	44
	3/0	0	0	1	1	2	4	6	9	12	16	26	37
	4/0	0	0	1	1	1	3	5	8	10	13	22	31
	250	0	0	1	1	1	2	4	6	8	11	17	25
	300	0	0	0	1	1	2	3	5	7	9	15	21
	350	0	0	0	1	1	1	3	5	6	8	13	19
	400	0	0	0	1	1	1	3	4	6	7	12	17
	500	0	0	0	1	1	1	2	3	5	6	10	14
	600	0	0	0	0	1	1	1	3	4	5	8	12
	700	0	0	0	0	1	1	1	2	3	4	7	10
	750	0	0	0	0	1	1	1	2	3	4	7	10
	900	0	0	0	0	0	1	1	2	3	4	6	8
	1000	0	0	0	0	0	1	1	1	2	3	5	8
THHN, THWN, THWN-2	8	—	—	—	—	—	—	—	—	—	—	—	—
	6	1	3	6	11	15	25	36	57	77	99	158	226
	4	1	1	3	6	9	15	22	35	47	61	98	140
	2	1	1	2	5	6	11	16	25	34	44	70	100
	1	1	1	1	3	5	8	12	19	25	33	53	75
	1/0	0	1	1	3	4	7	10	16	22	28	45	64
	2/0	0	1	1	2	3	6	8	13	18	23	37	53
	3/0	0	1	1	1	3	5	7	11	15	19	31	44
	4/0	0	0	1	1	2	4	6	9	12	16	25	37
	250	0	0	1	1	1	3	4	7	10	12	20	29
	300	0	0	1	1	1	3	4	6	8	11	17	25
	350	0	0	0	1	1	2	3	5	7	9	15	22
	400	0	0	0	1	1	1	3	5	6	8	13	19
	500	0	0	0	1	1	1	2	4	5	7	11	16
	600	0	0	0	1	1	1	1	3	4	6	9	13
	700	0	0	0	0	1	1	1	3	4	5	8	12
	750	0	0	0	0	1	1	1	3	4	5	8	11
	900	0	0	0	0	0	1	1	2	3	4	6	8
	1000	0	0	0	0	0	1	1	1	3	3	5	8
XHHW, XHHW-2	8	1	4	7	12	17	29	42	65	88	114	181	260
	6	1	3	5	9	13	21	31	48	65	85	134	193
	4	1	1	3	6	9	15	22	35	47	61	98	140
	2	1	1	2	5	6	11	16	25	34	44	70	100
	1	1	1	1	3	5	8	12	19	25	33	53	75
	1/0	0	1	1	3	4	7	10	16	22	28	45	64
	2/0	0	1	1	2	3	6	8	13	18	24	38	54
	3/0	0	1	1	1	3	5	7	11	15	19	31	44
	4/0	0	0	1	1	2	4	6	9	12	16	26	37
	250	0	0	1	1	1	3	5	7	10	13	21	30
	300	0	0	1	1	1	3	4	6	8	11	17	25
	350	0	0	1	1	1	2	3	5	7	10	15	22
	400	0	0	0	1	1	1	3	5	7	9	14	20
	500	0	0	0	1	1	1	2	4	5	7	11	17
	600	0	0	0	1	1	1	1	3	4	6	9	13
	700	0	0	0	0	1	1	1	3	4	5	8	12
	750	0	0	0	0	1	1	1	2	3	5	7	11
	900	0	0	0	0	1	1	1	2	3	4	6	8
	1000	0	0	0	0	0	1	1	1	3	3	6	8

Definition: *Compact stranding* is the result of a manufacturing process where the standard conductor is compressed to the extent that the interstices (voids between strand wires) are virtually eliminated.

Table C.10 Maximum Number of Conductors or Fixture Wires in Rigid PVC Conduit, Schedule 40 and HDPE Conduit (*Based on Table 1, Chapter 9*)

		CONDUCTORS											
	Conductor Size	Metric Designator (Trade Size)											
Type	(AWG/ kcmil)	16 (½)	21 (¾)	27 (1)	35 (1¼)	41 (1½)	53 (2)	63 (2½)	78 (3)	91 (3½)	103 (4)	129 (5)	155 (6)
RHH,	14	4	7	11	20	27	45	64	99	133	171	269	390
RHW,	12	3	5	9	16	22	37	53	82	110	142	224	323
RHW-2	10	2	4	7	13	18	30	43	66	89	115	181	261
	8	1	2	4	7	9	15	22	35	46	60	94	137
	6	1	1	3	5	7	12	18	28	37	48	76	109
	4	1	1	2	4	6	10	14	22	29	37	59	85
	3	1	1	1	4	5	8	12	19	25	33	52	75
	2	1	1	1	3	4	7	10	16	22	28	45	65
	1	0	1	1	1	3	5	7	11	14	19	29	43
	1/0	0	1	1	1	2	4	6	9	13	16	26	37
	2/0	0	0	1	1	1	3	5	8	11	14	22	32
	3/0	0	0	1	1	1	3	4	7	9	12	19	28
	4/0	0	0	1	1	1	2	4	6	8	10	16	24
	250	0	0	0	1	1	1	3	4	6	8	12	18
	300	0	0	0	1	1	1	2	4	5	7	11	16
	350	0	0	0	1	1	1	2	3	5	6	10	14
	400	0	0	0	1	1	1	1	3	4	6	9	13
	500	0	0	0	0	1	1	1	3	4	5	8	11
	600	0	0	0	0	1	1	1	2	3	4	6	9
	700	0	0	0	0	0	1	1	1	3	3	6	8
	750	0	0	0	0	0	1	1	1	2	3	5	8
	800	0	0	0	0	0	1	1	1	2	3	5	7
	900	0	0	0	0	0	1	1	1	2	3	5	7
	1000	0	0	0	0	0	1	1	1	1	3	4	6
	1250	0	0	0	0	0	0	1	1	1	1	3	5
	1500	0	0	0	0	0	0	1	1	1	1	3	4
	1750	0	0	0	0	0	0	1	1	1	1	2	3
	2000	0	0	0	0	0	0	1	1	1	1	2	3
TW	14	8	14	24	42	57	94	135	209	280	361	568	822
	12	6	11	18	32	44	72	103	160	215	277	436	631
	10	4	8	13	24	32	54	77	119	160	206	325	470
	8	2	4	7	13	18	30	43	66	89	115	181	261
RHH*, RHW*, RHW-2*, THHW, THW, THW-2	14	5	9	16	28	38	63	90	139	186	240	378	546
RHH*, RHW*, RHW-2*, THHW, THW	12	4	8	12	22	30	50	72	112	150	193	304	439
	10	3	6	10	17	24	39	56	87	117	150	237	343
RHH*, RHW*, RHW-2*, THHW, THW, THW-2	8	1	3	6	10	14	23	33	52	70	90	142	205

(Continues)

Table C.10 *Continued*

		CONDUCTORS											
	Conductor Size	Metric Designator (Trade Size)											
Type	(AWG/ kcmil)	16 (½)	21 (¾)	27 (1)	35 (1¼)	41 (1½)	53 (2)	63 (2½)	78 (3)	91 (3½)	103 (4)	129 (5)	155 (6)
RHH*,	6	1	2	4	8	11	18	26	40	53	69	109	157
RHW*,	4	1	1	3	6	8	13	19	30	40	51	81	117
RHW-2*	3	1	1	3	5	7	11	16	25	34	44	69	100
TW,	2	1	1	2	4	6	10	14	22	29	37	59	85
THW,	1	0	1	1	3	4	7	10	15	20	26	41	60
THHW,	1/0	0	1	1	2	3	6	8	13	17	22	35	51
THW-2	2/0	0	1	1	1	3	5	7	11	15	19	30	43
	3/0	0	1	1	1	2	4	6	9	12	16	25	36
	4/0	0	0	1	1	1	3	5	8	10	13	21	30
	250	0	0	1	1	1	3	4	6	8	11	17	25
	300	0	0	1	1	1	2	3	5	7	9	15	21
	350	0	0	0	1	1	1	3	5	6	8	13	19
	400	0	0	0	1	1	1	3	4	6	7	12	17
	500	0	0	0	1	1	1	2	3	5	6	10	14
	600	0	0	0	0	1	1	1	3	4	5	8	11
	700	0	0	0	0	1	1	1	2	3	4	7	10
	750	0	0	0	0	1	1	1	2	3	4	6	10
	800	0	0	0	0	1	1	1	2	3	4	6	9
	900	0	0	0	0	0	1	1	1	3	3	6	8
	1000	0	0	0	0	0	1	1	1	2	3	5	7
	1250	0	0	0	0	0	1	1	1	1	2	4	6
	1500	0	0	0	0	0	1	1	1	1	1	3	5
	1750	0	0	0	0	0	0	1	1	1	1	3	4
	2000	0	0	0	0	0	0	1	1	1	1	3	4
THHN,	14	11	21	34	60	82	135	193	299	401	517	815	1178
THWN,	12	8	15	25	43	59	99	141	218	293	377	594	859
THWN-2	10	5	9	15	27	37	62	89	137	184	238	374	541
	8	3	5	9	16	21	36	51	79	106	137	216	312
	6	1	4	6	11	15	26	37	57	77	99	156	225
	4	1	2	4	7	9	16	22	35	47	61	96	138
	3	1	1	3	6	8	13	19	30	40	51	81	117
	2	1	1	3	5	7	11	16	25	33	43	68	98
	1	1	1	1	3	5	8	12	18	25	32	50	73
	1/0	1	1	1	3	4	7	10	15	21	27	42	61
	2/0	0	1	1	2	3	6	8	13	17	22	35	51
	3/0	0	1	1	1	3	5	7	11	14	18	29	42
	4/0	0	1	1	1	2	4	6	9	12	15	24	35
	250	0	0	1	1	1	3	4	7	10	12	20	28
	300	0	0	1	1	1	3	4	6	8	11	17	24
	350	0	0	1	1	1	2	3	5	7	9	15	21
	400	0	0	0	1	1	1	3	5	6	8	13	19
	500	0	0	0	1	1	1	2	4	5	7	11	16
	600	0	0	0	1	1	1	1	3	4	5	9	13
	700	0	0	0	0	1	1	1	3	4	5	8	11
	750	0	0	0	0	1	1	1	2	3	4	7	11
	800	0	0	0	0	1	1	1	2	3	4	7	10
	900	0	0	0	0	1	1	1	2	3	4	6	9
	1000	0	0	0	0	0	1	1	1	3	3	6	8
FEP,	14	11	20	33	58	79	131	188	290	389	502	790	1142
FEPB,	12	8	15	24	42	58	96	137	212	284	366	577	834
PFA,	10	6	10	17	30	41	69	98	152	204	263	414	598
PFAH,	8	3	6	10	17	24	39	56	87	117	150	237	343
TFE	6	2	4	7	12	17	28	40	62	83	107	169	244
	4	1	3	5	8	12	19	28	43	58	75	118	170
	3	1	2	4	7	10	16	23	36	48	62	98	142
	2	1	1	3	6	8	13	19	30	40	51	81	117

Table C.10 *Continued*

		CONDUCTORS											
	Conductor Size	Metric Designator (Trade Size)											
Type	(AWG/ kcmil)	16 (½)	21 (¾)	27 (1)	35 (1¼)	41 (1½)	53 (2)	63 (2½)	78 (3)	91 (3½)	103 (4)	129 (5)	155 (6)
PFA, PFAH, TFE	1	1	1	2	4	5	9	13	20	28	36	56	81
PFA, PFAH, TFE, Z	1/0	1	1	1	3	4	8	11	17	23	30	47	68
	2/0	0	1	1	3	4	6	9	14	19	24	39	56
	3/0	0	1	1	2	3	5	7	12	16	20	32	46
	4/0	0	1	1	1	2	4	6	9	13	16	26	38
Z	14	13	24	40	70	95	158	226	350	469	605	952	1376
	12	9	17	28	49	68	112	160	248	333	429	675	976
	10	6	10	17	30	41	69	98	152	204	263	414	598
	8	3	6	11	19	26	43	62	96	129	166	261	378
	6	2	4	7	13	18	30	43	67	90	116	184	265
	4	1	3	5	9	12	21	30	46	62	80	126	183
	3	1	2	4	6	9	15	22	34	45	58	92	133
	2	1	1	3	5	7	12	18	28	38	49	77	111
	1	1	1	2	4	6	10	14	23	30	39	62	90
XHH, XHHW, XHHW-2 ZW	14	8	14	24	42	57	94	135	209	280	361	568	822
	12	6	11	18	32	44	72	103	160	215	277	436	631
	10	4	8	13	24	32	54	77	119	160	206	325	470
	8	2	4	7	13	18	30	43	66	89	115	181	261
	6	1	3	5	10	13	22	32	49	66	85	134	193
	4	1	2	4	7	9	16	23	35	48	61	97	140
	3	1	1	3	6	8	13	19	30	40	52	82	118
	2	1	1	3	5	7	11	16	25	34	44	69	99
XHH, XHHW, XHHW-2	1	1	1	1	3	5	8	12	19	25	32	51	74
	1/0	1	1	1	3	4	7	10	16	21	27	43	62
	2/0	0	1	1	2	3	6	8	13	17	23	36	52
	3/0	0	1	1	1	3	5	7	11	14	19	30	43
	4/0	0	1	1	1	2	4	6	9	12	15	24	35
	250	0	0	1	1	1	3	5	7	10	13	20	29
	300	0	0	1	1	1	3	4	6	8	11	17	25
	350	0	0	1	1	1	2	3	5	7	9	15	22
	400	0	0	0	1	1	1	3	5	6	8	13	19
	500	0	0	0	1	1	1	2	4	5	7	11	16
	600	0	0	0	1	1	1	1	3	4	5	9	13
	700	0	0	0	0	1	1	1	3	4	5	8	11
	750	0	0	0	0	1	1	1	2	3	4	7	11
	800	0	0	0	0	1	1	1	2	3	4	7	10
	900	0	0	0	0	1	1	1	2	3	4	6	9
	1000	0	0	0	0	0	1	1	1	3	3	6	8
	1250	0	0	0	0	0	1	1	1	1	3	4	6
	1500	0	0	0	0	0	1	1	1	1	2	4	5
	1750	0	0	0	0	0	0	1	1	1	1	3	5
	2000	0	0	0	0	0	0	1	1	1	1	3	4

(Continues)

Table C.10 *Continued*

		FIXTURE WIRES					
	Conductor Size (AWG/ kcmil)	Metric Designator (Trade Size)					
Type		16 (½)	21 (¾)	27 (1)	35 (1¼)	41 (1½)	53 (2)
FFH-2, RFH-2, RFHH-3	18	8	14	23	40	54	90
	16	6	12	19	33	46	76
SF-2, SFF-2	18	10	17	29	50	69	114
	16	8	14	24	42	57	94
	14	6	12	19	33	46	76
SF-1, SFF-1	18	17	31	51	89	122	202
RFHH-2, TF, TFF, XF, XFF RFH-1,	18	13	23	38	66	90	149
RFHH-2, TF, TFF, XF, XFF	16	10	18	30	53	73	120
XF, XFF	14	8	14	24	42	57	94
TFN, TFFN	18	20	37	60	105	144	239
	16	16	28	46	80	110	183
PF, PFF, PGF, PGFF, PAF, PTF, PTFF, PAFF	18	19	35	57	100	137	227
	16	15	27	44	77	106	175
	14	11	20	33	58	79	131
HF, HFF, ZF, ZFF, ZHF	18	25	45	74	129	176	292
	16	18	33	54	95	130	216
	14	13	24	40	70	95	158
KF-2, KFF-2	18	36	65	107	187	256	424
	16	26	46	75	132	180	299
	14	17	31	52	90	124	205
	12	12	22	35	62	85	141
	10	8	14	24	42	57	94
KF-1, KFF-1	18	43	78	128	223	305	506
	16	30	55	90	157	214	355
	14	20	37	60	105	144	239
	12	13	24	40	70	95	158
	10	9	16	26	45	62	103
XF, XFF	12	4	8	12	22	30	50
	10	3	6	10	17	24	39

Notes:

1. This table is for concentric stranded conductors only. For compact stranded conductors, Table C.10(A) should be used.

2. Two-hour fire-rated RHH cable has ceramifiable insulation which has much larger diameters than other RHH wires. Consult manufacturer's conduit fill tables.

*Types RHH, RHW, and RHW-2 without outer covering.

Table C.10(A) Maximum Number of Compact Conductors in Rigid PVC Conduit, Schedule 40 and HDPE Conduit (*Based on Table 1, Chapter 9*)

		COMPACT CONDUCTORS										
	Conductor Size	Metric Designator (Trade Size)										
Type	(AWG/ kcmil)	16 (½)	21 (¾)	27 (1)	35 (1¼)	41 (1½)	53 (2)	63 (2½)	78 (3)	91 (3½)	103 (4)	129 (5)	155 (6)
THW,	8	1	4	6	11	15	26	37	57	76	98	155	224
THW-2,	6	1	3	5	9	12	20	28	44	59	76	119	173
THHW	4	1	1	3	6	9	15	21	33	44	57	89	129
	2	1	1	2	5	6	11	15	24	32	42	66	95
	1	1	1	1	3	4	7	11	17	23	29	46	67
	1/0	0	1	1	3	4	6	9	15	20	25	40	58
	2/0	0	1	1	2	3	5	8	12	16	21	34	49
	3/0	0	1	1	1	3	5	7	10	14	18	29	42
	4/0	0	1	1	1	2	4	5	9	12	15	24	35
	250	0	0	1	1	1	3	4	7	9	12	19	27
	300	0	0	1	1	1	2	4	6	8	10	16	24
	350	0	0	1	1	1	2	3	5	7	9	15	21
	400	0	0	0	1	1	1	3	5	6	8	13	19
	500	0	0	0	1	1	1	2	4	5	7	11	16
	600	0	0	0	1	1	1	1	3	4	5	9	13
	700	0	0	0	0	1	1	1	3	4	5	8	12
	750	0	0	0	0	1	1	1	2	3	5	7	11
	900	0	0	0	0	1	1	1	2	3	4	6	9
	1000	0	0	0	0	1	1	1	1	3	4	6	9
THHN,	8	—	—	—	—	—	—	—	—	—	—	—	—
THWN,	6	2	4	7	13	17	29	41	64	86	111	175	253
THWN-2	4	1	2	4	8	11	18	25	40	53	68	108	156
	2	1	1	3	5	8	13	18	28	38	49	77	112
	1	1	1	2	4	6	9	14	21	29	37	58	84
	1/0	1	1	1	3	5	8	12	18	24	31	49	72
	2/0	0	1	1	3	4	7	9	15	20	26	41	59
	3/0	0	1	1	2	3	5	8	12	17	22	34	50
	4/0	0	1	1	1	3	4	6	10	14	18	28	41
	250	0	0	1	1	1	3	5	8	11	14	22	32
	300	0	0	1	1	1	3	4	7	9	12	19	28
	350	0	0	1	1	1	3	4	6	8	10	17	24
	400	0	0	1	1	1	2	3	5	7	9	15	22
	500	0	0	0	1	1	1	3	4	6	8	13	18
	600	0	0	0	1	1	1	2	4	5	6	10	15
	700	0	0	0	1	1	1	1	3	4	5	9	13
	750	0	0	0	1	1	1	1	3	4	5	8	12
	900	0	0	0	0	1	1	1	2	3	4	6	9
	1000	0	0	0	0	1	1	1	2	3	4	6	9
XHHW,	8	3	5	8	14	20	33	47	73	99	127	200	290
XHHW-2	6	1	4	6	11	15	25	35	55	73	94	149	215
	4	1	2	4	8	11	18	25	40	53	68	108	156
	2	1	1	3	5	8	13	18	28	38	49	77	112
	1	1	1	2	4	6	9	14	21	29	37	58	84
	1/0	1	1	1	3	5	8	12	18	24	31	49	72
	2/0	1	1	1	3	4	7	10	15	20	26	42	60
	3/0	0	1	1	2	3	5	8	12	17	22	34	50
	4/0	0	1	1	1	3	5	7	10	14	18	29	42
	250	0	0	1	1	1	4	5	8	11	14	23	33
	300	0	0	1	1	1	3	4	7	9	12	19	28
	350	0	0	1	1	1	3	4	6	8	11	17	25
	400	0	0	1	1	1	2	3	5	7	10	15	22
	500	0	0	0	1	1	1	3	4	6	8	13	18
	600	0	0	0	1	1	1	2	4	5	6	10	15
	700	0	0	0	1	1	1	1	3	4	5	9	13
	750	0	0	0	1	1	1	1	3	4	5	8	12
	900	0	0	0	0	1	1	1	2	3	4	6	9
	1000	0	0	0	0	1	1	1	2	3	4	6	9

Definition: *Compact stranding* is the result of a manufacturing process where the standard conductor is compressed to the extent that the interstices (voids between strand wires) are virtually eliminated.

Table C.11 Maximum Number of Conductors or Fixture Wires in Type A, Rigid PVC Conduit (*Based on Table 1, Chapter 9*)

		CONDUCTORS									
	Conductor Size	Metric Designator (Trade Size)									
Type	(AWG/ kcmil)	16 (½)	21 (¾)	27 (1)	35 (1¼)	41 (1½)	53 (2)	63 (2½)	78 (3)	91 (3½)	103 (4)
RHH,	14	5	9	15	24	31	49	74	112	146	187
RHW,	12	4	7	12	20	26	41	61	93	121	155
RHW-2	10	3	6	10	16	21	33	50	75	98	125
	8	1	3	5	8	11	17	26	39	51	65
	6	1	2	4	6	9	14	21	31	41	52
	4	1	1	3	5	7	11	16	24	32	41
	3	1	1	3	4	6	9	14	21	28	36
	2	1	1	2	4	5	8	12	18	24	31
	1	0	1	1	2	3	5	8	12	16	20
	1/0	0	1	1	2	3	5	7	10	14	18
	2/0	0	1	1	1	2	4	6	9	12	15
	3/0	0	1	1	1	1	3	5	8	10	13
	4/0	0	0	1	1	1	3	4	7	9	11
	250	0	0	1	1	1	1	3	5	7	8
	300	0	0	1	1	1	1	3	4	6	7
	350	0	0	0	1	1	1	2	4	5	7
	400	0	0	0	1	1	1	2	4	5	6
	500	0	0	0	1	1	1	1	3	4	5
	600	0	0	0	0	1	1	1	2	3	4
	700	0	0	0	0	1	1	1	2	3	4
	750	0	0	0	0	1	1	1	1	3	4
	800	0	0	0	0	1	1	1	1	3	3
	900	0	0	0	0	0	1	1	1	2	3
	1000	0	0	0	0	0	1	1	1	2	3
	1250	0	0	0	0	0	1	1	1	1	2
	1500	0	0	0	0	0	0	1	1	1	1
	1750	0	0	0	0	0	0	1	1	1	1
	2000	0	0	0	0	0	0	1	1	1	1
TW	14	11	18	31	51	67	105	157	235	307	395
	12	8	14	24	39	51	80	120	181	236	303
	10	6	10	18	29	38	60	89	135	176	226
	8	3	6	10	16	21	33	50	75	98	125
RHH*, RHW*, RHW-2*, THHW, THW, THW-2	14	7	12	20	34	44	70	104	157	204	262
RHH*, RHW*, RHW-2*, THHW, THW	12	6	10	16	27	35	56	84	126	164	211
	10	4	8	13	21	28	44	65	98	128	165
RHH*, RHW*, RHW-2*, THHW, THW, THW-2	8	2	4	8	12	16	26	39	59	77	98
RHH*, RHW*, RHW-2*, TW, THHW, THW, THW-2	6	1	3	6	9	13	20	30	45	59	75

Table C.11 *Continued*

		CONDUCTORS								
	Conductor Size	Metric Designator (Trade Size)								
Type	(AWG/ kcmil)	16 (½)	21 (¾)	27 (1)	35 (1¼)	41 (1½)	53 (2)	63 (2½)	78 (3)	91 (3½)	103 (4)
RHH*,	4	1	2	4	7	9	15	22	33	44	56
RHW*,	3	1	1	4	6	8	13	19	29	37	48
RHW-2*,	2	1	1	3	5	7	11	16	24	32	41
TW,	1	1	1	1	3	5	7	11	17	22	29
THW, THHW,	1/0	1	1	1	3	4	6	10	14	19	24
THW-2	2/0	0	1	1	2	3	5	8	12	16	21
	3/0	0	1	1	1	3	4	7	10	13	17
	4/0	0	1	1	1	2	4	6	9	11	14
	250	0	0	1	1	1	3	4	7	9	12
	300	0	0	1	1	1	2	4	6	8	10
	350	0	0	1	1	1	2	3	5	7	9
	400	0	0	1	1	1	1	3	5	6	8
	500	0	0	0	1	1	1	2	4	5	7
	600	0	0	0	1	1	1	1	3	4	5
	700	0	0	0	1	1	1	1	3	4	5
	750	0	0	0	1	1	1	1	3	3	4
	800	0	0	0	0	1	1	1	2	3	4
	900	0	0	0	0	1	1	1	2	3	4
	1000	0	0	0	0	1	1	1	1	3	3
	1250	0	0	0	0	0	1	1	1	1	3
	1500	0	0	0	0	0	1	1	1	1	2
	1750	0	0	0	0	0	0	1	1	1	1
	2000	0	0	0	0	0	0	1	1	1	1
THHN,	14	16	27	44	73	96	150	225	338	441	566
THWN,	12	11	19	32	53	70	109	164	246	321	412
THWN-2	10	7	12	20	33	44	69	103	155	202	260
	8	4	7	12	19	25	40	59	89	117	150
	6	3	5	8	14	18	28	43	64	84	108
	4	1	3	5	8	11	17	26	39	52	66
	3	1	2	4	7	9	15	22	33	44	56
	2	1	1	3	6	8	12	19	28	37	47
	1	1	1	2	4	6	9	14	21	27	35
	1/0	1	1	2	4	5	8	11	17	23	29
	2/0	1	1	1	3	4	6	10	14	19	24
	3/0	0	1	1	2	3	5	8	12	16	20
	4/0	0	1	1	1	3	4	6	10	13	17
	250	0	1	1	1	2	3	5	8	10	14
	300	0	0	1	1	1	3	4	7	9	12
	350	0	0	1	1	1	2	4	6	8	10
	400	0	0	1	1	1	2	3	5	7	9
	500	0	0	1	1	1	1	3	4	6	7
	600	0	0	0	1	1	1	2	3	5	6
	700	0	0	0	1	1	1	1	3	4	5
	750	0	0	0	1	1	1	1	3	4	5
	800	0	0	0	1	1	1	1	3	4	5
	900	0	0	0	0	1	1	1	2	3	4
	1000	0	0	0	0	1	1	1	2	3	4
FEP,	14	15	26	43	70	93	146	218	327	427	549
FEPB,	12	11	19	31	51	68	106	159	239	312	400
PFA,	10	8	13	22	37	48	76	114	171	224	287
PFAH,	8	4	8	13	21	28	44	65	98	128	165
TFE	6	3	5	9	15	20	31	46	70	91	117
	4	1	4	6	10	14	21	32	49	64	82
	3	1	3	5	8	11	18	27	40	53	68
	2	1	2	4	7	9	15	22	33	44	56

(Continues)

Table C.11 *Continued*

		CONDUCTORS									
	Conductor Size	Metric Designator (Trade Size)									
Type	(AWG/ kcmil)	16 (½)	21 (¾)	27 (1)	35 (1¼)	41 (1½)	53 (2)	63 (2½)	78 (3)	91 (3½)	103 (4)
PFA, PFAH, TFE	1	1	1	3	5	6	10	15	23	30	39
PFA, PFAH, TFE, Z	1/0	1	1	2	4	5	8	13	19	25	32
	2/0	1	1	1	3	4	7	10	16	21	27
	3/0	1	1	1	3	3	6	9	13	17	22
	4/0	0	1	1	2	3	5	7	11	14	18
Z	14	18	31	52	85	112	175	263	395	515	661
	12	13	22	37	60	79	124	186	280	365	469
	10	8	13	22	37	48	76	114	171	224	287
	8	5	8	14	23	30	48	72	108	141	181
	6	3	6	10	16	21	34	50	76	99	127
	4	2	4	7	11	15	23	35	52	68	88
	3	1	3	5	8	11	17	25	38	50	64
	2	1	2	4	7	9	14	21	32	41	53
	1	1	1	3	5	7	11	17	26	33	43
XHH, XHHW, XHHW-2, ZW	14	11	18	31	51	67	105	157	235	307	395
	12	8	14	24	39	51	80	120	181	236	303
	10	6	10	18	29	38	60	89	135	176	226
	8	3	6	10	16	21	33	50	75	98	125
	6	2	4	7	12	15	24	37	55	72	93
	4	1	3	5	8	11	18	26	40	52	67
	3	1	2	4	7	9	15	22	34	44	57
	2	1	1	3	6	8	12	19	28	37	48
XHH, XHHW, XHHW-2	1	1	1	3	4	6	9	14	21	28	35
	1/0	1	1	2	4	5	8	12	18	23	30
	2/0	1	1	1	3	4	6	10	15	19	25
	3/0	0	1	1	2	3	5	8	12	16	20
	4/0	0	1	1	1	3	4	7	10	13	17
	250	0	1	1	1	2	3	5	8	11	14
	300	0	0	1	1	1	3	5	7	9	12
	350	0	0	1	1	1	3	4	6	8	10
	400	0	0	1	1	1	2	3	5	7	9
	500	0	0	1	1	1	1	3	4	6	8
	600	0	0	0	1	1	1	2	3	5	6
	700	0	0	0	1	1	1	1	3	4	5
	750	0	0	0	1	1	1	1	3	4	5
	800	0	0	0	1	1	1	1	3	4	5
	900	0	0	0	0	1	1	1	2	3	4
	1000	0	0	0	0	1	1	1	2	3	4
	1250	0	0	0	0	0	1	1	1	2	3
	1500	0	0	0	0	0	1	1	1	1	2
	1750	0	0	0	0	0	1	1	1	1	2
	2000	0	0	0	0	0	0	1	1	1	1

Table C.11 *Continued*

	Conductor Size (AWG/ kcmil)	Metric Designator (Trade Size)					
Type		**16** **(½)**	**21** **(¾)**	**27** **(1)**	**35** **(1¼)**	**41** **(1½)**	**53** **(2)**
FFH-2,	18	10	18	30	48	64	100
RFH-2, RFHH-3	16	9	15	25	41	54	85
SF-2, SFF-2	18	13	22	37	61	81	127
	16	11	18	31	51	67	105
	14	9	15	25	41	54	85
SF-1, SFF-1	18	23	40	66	108	143	224
RFH-1, RFHH-2, TF, TFF, XF, XFF	18	17	29	49	80	105	165
RFHH-2, TF, TFF, XF, XFF	16	14	24	39	65	85	134
XF, XFF	14	11	18	31	51	67	105
TFN, TFFN	18	28	47	79	128	169	265
	16	21	36	60	98	129	202
PF, PFF, PGF, PGFF, PAF, PTF, PTFF, PAFF	18	26	45	74	122	160	251
	16	20	34	58	94	124	194
	14	15	26	43	70	93	146
HF, HFF, ZF, ZFF, ZHF	18	34	58	96	157	206	324
	16	25	42	71	116	152	239
	14	18	31	52	85	112	175
KF-2, KFF-2	18	49	84	140	228	300	470
	16	35	59	98	160	211	331
	14	24	40	67	110	145	228
	12	16	28	46	76	100	157
	10	11	18	31	51	67	105
KF-1, KFF-1	18	59	100	167	272	357	561
	16	41	70	117	191	251	394
	14	28	47	79	128	169	265
	12	18	31	52	85	112	175
	10	12	20	34	55	73	115
XF, XFF	12	6	10	16	27	35	56
	10	4	8	13	21	28	44

Notes:

1. This table is for concentric stranded conductors only. For compact stranded conductors, Table C.11(A) should be used.

2. Two-hour fire-rated RHH cable has ceramifiable insulation which has much larger diameters than other RHH wires. Consult manufacturer's conduit fill tables.

*Types RHH, RHW, and RWH-2 without outer covering.

Table C.11(A) Maximum Number of Compact Conductors in Type A, Rigid PVC Conduit
(Based on Table 1, Chapter 9)

		COMPACT CONDUCTORS									
	Conductor Size	Metric Designator (Trade Size)									
Type	(AWG/ kcmil)	16 (½)	21 (¾)	27 (1)	35 (1¼)	41 (1½)	53 (2)	63 (2½)	78 (3)	91 (3½)	103 (4)
THW,	8	3	5	8	14	18	28	42	64	84	107
THW-2,	6	2	4	6	10	14	22	33	49	65	83
THHW	4	1	3	5	8	10	16	24	37	48	62
	2	1	1	3	6	7	12	18	27	36	46
	1	1	1	2	4	5	8	13	19	25	32
	1/0	1	1	1	3	4	7	11	16	21	28
	2/0	1	1	1	3	4	6	9	14	18	23
	3/0	0	1	1	2	3	5	8	12	15	20
	4/0	0	1	1	1	3	4	6	10	13	17
	250	0	1	1	1	1	3	5	8	10	13
	300	0	0	1	1	1	3	4	7	9	11
	350	0	0	1	1	1	2	4	6	8	10
	400	0	0	1	1	1	2	3	5	7	9
	500	0	0	1	1	1	1	3	4	6	8
	600	0	0	0	1	1	1	2	3	5	6
	700	0	0	0	1	1	1	1	3	4	5
	750	0	0	0	1	1	1	1	3	4	5
	900	0	0	0	0	1	1	2	2	3	4
	1000	0	0	0	0	1	1	1	2	3	4
THHN,	8	—	—	—	—	—	—	—	—	—	—
THWN,	6	3	5	9	15	20	32	48	72	94	121
THWN-2	4	1	3	6	9	12	20	30	45	58	75
	2	1	2	4	7	9	14	21	32	42	54
	1	1	1	3	5	7	10	16	24	31	40
	1/0	1	1	2	4	6	9	13	20	27	34
	2/0	1	1	1	3	5	7	11	17	22	28
	3/0	1	1	1	3	4	6	9	14	18	24
	4/0	0	1	1	2	3	5	8	11	15	19
	250	0	1	1	1	2	4	6	9	12	15
	300	0	1	1	1	1	3	5	8	10	13
	350	0	0	1	1	1	3	4	7	9	11
	400	0	0	1	1	1	2	4	6	8	10
	500	0	0	1	1	1	2	3	5	7	9
	600	0	0	0	1	1	1	3	4	5	7
	700	0	0	0	1	1	1	2	3	5	6
	750	0	0	0	1	1	1	2	3	4	6
	900	0	0	0	1	1	1	2	3	4	5
	1000	0	0	0	0	1	1	1	2	3	4
XHHW,	8	4	6	11	18	23	37	55	83	108	139
XHHW-2	6	3	5	8	13	17	27	41	62	80	103
	4	1	3	6	9	12	20	30	45	58	75
	2	1	2	4	7	9	14	21	32	42	54
	1	1	1	3	5	7	10	16	24	31	40
	1/0	1	1	2	4	6	9	13	20	27	34
	2/0	1	1	1	3	5	7	11	17	22	29
	3/0	1	1	1	3	4	6	9	14	18	24
	4/0	0	1	1	2	3	5	8	12	15	20
	250	0	1	1	1	2	4	6	9	12	16
	300	0	1	1	1	1	3	5	8	10	13
	350	0	0	1	1	1	3	5	7	9	12
	400	0	0	1	1	1	3	4	6	8	11
	500	0	0	1	1	1	2	3	5	7	9
	600	0	0	0	1	1	1	3	4	5	7
	700	0	0	0	1	1	1	2	3	5	6
	750	0	0	0	1	1	1	2	3	4	6
	900	0	0	0	1	1	1	2	3	4	5
	1000	0	0	0	0	1	1	1	2	3	4

Definition: *Compact stranding* is the result of a manufacturing process where the standard conductor is compressed to the extent that the interstices (voids between strand wires) are virtually eliminated.

Table C.12 Maximum Number of Conductors in Type EB, PVC Conduit
(Based on Table 1, Chapter 9)

		CONDUCTORS					
	Conductor Size	Metric Designator (Trade Size)					
Type	(AWG/ kcmil)	53 (2)	78 (3)	91 (3½)	103 (4)	129 (5)	155 (6)
RHH, RHW, RHW-2	14	53	119	155	197	303	430
	12	44	98	128	163	251	357
	10	35	79	104	132	203	288
	8	18	41	54	69	106	151
	6	15	33	43	55	85	121
	4	11	26	34	43	66	94
	3	10	23	30	38	58	83
	2	9	20	26	33	50	72
	1	6	13	17	21	33	47
	1/0	5	11	15	19	29	41
	2/0	4	10	13	16	25	36
	3/0	4	8	11	14	22	31
	4/0	3	7	9	12	18	26
	250	2	5	7	9	14	20
	300	1	5	6	8	12	17
	350	1	4	5	7	11	16
	400	1	4	5	6	10	14
	500	1	3	4	5	9	12
	600	1	3	3	4	7	10
	700	1	2	3	4	6	9
	750	1	2	3	4	6	9
	800	1	2	3	4	6	8
	900	1	1	2	3	5	7
	1000	1	1	2	3	5	7
	1250	1	1	1	2	3	5
	1500	0	1	1	1	3	4
	1750	0	1	1	1	3	4
	2000	0	1	1	1	2	3
TW	14	111	250	327	415	638	907
	12	85	192	251	319	490	696
	10	63	143	187	238	365	519
	8	35	79	104	132	203	288
RHH*, RHW*, RHW-2*, THHW, THW, THW-2	14	74	166	217	276	424	603
RHH*, RHW*, RHW-2*, THHW, THW	12	59	134	175	222	341	485
	10	46	104	136	173	266	378
RHH*, RHW*, RHW-2*, THHW, THW, THW-2	8	28	62	81	104	159	227

(Continues)

Table C.12 *Continued*

	Conductor Size (AWG/ kcmil)	\multicolumn{6}{c}{CONDUCTORS}

		\multicolumn{6}{c}{Metric Designator (Trade Size)}
Type		53 (2)	78 (3)	91 (3½)	103 (4)	129 (5)	155 (6)
RHH*,	6	21	48	62	79	122	173
RHW*,	4	16	36	46	59	91	129
RHW-2*,	3	13	30	40	51	78	111
TW, THW,	2	11	26	34	43	66	94
THHW,	1	8	18	24	30	46	66
THW-2	1/0	7	15	20	26	40	56
	2/0	6	13	17	22	34	48
	3/0	5	11	14	18	28	40
	4/0	4	9	12	15	24	34
	250	3	7	10	12	19	27
	300	3	6	8	11	17	24
	350	2	6	7	9	15	21
	400	2	5	7	8	13	19
	500	1	4	5	7	11	16
	600	1	3	4	6	9	13
	700	1	3	4	5	8	11
	750	1	3	4	5	7	11
	800	1	3	3	4	7	10
	900	1	2	3	4	6	9
	1000	1	2	3	4	6	8
	1250	1	1	2	3	4	6
	1500	1	1	1	2	4	6
	1750	1	1	1	2	3	5
	2000	0	1	1	1	3	4
THHN,	14	159	359	468	595	915	1300
THWN,	12	116	262	342	434	667	948
THWN-2	10	73	165	215	274	420	597
	8	42	95	124	158	242	344
	6	30	68	89	114	175	248
	4	19	42	55	70	107	153
	3	16	36	46	59	91	129
	2	13	30	39	50	76	109
	1	10	22	29	37	57	80
	1/0	8	18	24	31	48	68
	2/0	7	15	20	26	40	56
	3/0	5	13	17	21	33	47
	4/0	4	10	14	18	27	39
	250	4	8	11	14	22	31
	300	3	7	10	12	19	27
	350	3	6	8	11	17	24
	400	2	6	7	10	15	21
	500	1	5	6	8	12	18
	600	1	4	5	6	10	14
	700	1	3	4	6	9	12
	750	1	3	4	5	8	12
	800	1	3	4	5	8	11
	900	1	3	3	4	7	10
	1000	1	2	3	4	6	9
FEP, FEPB,	14	155	348	454	578	888	1261
PFA, PFAH,	12	113	254	332	422	648	920
TFE	10	81	182	238	302	465	660
	8	46	104	136	173	266	378
	6	33	74	97	123	189	269
	4	23	52	68	86	132	188
	3	19	43	56	72	110	157
	2	16	36	46	59	91	129

Table C.12 *Continued*

		CONDUCTORS					
	Conductor Size	Metric Designator (Trade Size)					
Type	(AWG/ kcmil)	53 (2)	78 (3)	91 (3½)	103 (4)	129 (5)	155 (6)
PFA, PFAH, TFE	1	11	25	32	41	63	90
PFA, PFAH, TFE, Z	1/0	9	20	27	34	53	75
	2/0	7	17	22	28	43	62
	3/0	6	14	18	23	36	51
	4/0	5	11	15	19	29	42
Z	14	186	419	547	696	1069	1519
	12	132	297	388	494	759	1078
	10	81	182	238	302	465	660
	8	51	115	150	191	294	417
	6	36	81	105	134	206	293
	4	24	55	72	92	142	201
	3	18	40	53	67	104	147
	2	15	34	44	56	86	122
	1	12	27	36	45	70	99
XHH, XHHW, XHHW-2, ZW	14	111	250	327	415	638	907
	12	85	192	251	319	490	696
	10	63	143	187	238	365	519
	8	35	79	104	132	203	288
	6	26	59	77	98	150	213
	4	19	42	56	71	109	155
	3	16	36	47	60	92	131
	2	13	30	39	50	77	110
XHH, XHHW, XHHW-2	1	10	22	29	37	58	82
	1/0	8	19	25	31	48	69
	2/0	7	16	20	26	40	57
	3/0	6	13	17	22	33	47
	4/0	5	11	14	18	27	39
	250	4	9	11	15	22	32
	300	3	7	10	12	19	28
	350	3	6	9	11	17	24
	400	2	6	8	10	15	22
	500	1	5	6	8	12	18
	600	1	4	5	6	10	14
	700	1	3	4	6	9	12
	750	1	3	4	5	8	12
	800	1	3	4	5	8	11
	900	1	3	3	4	7	10
	1000	1	2	3	4	6	9
	1250	1	1	2	3	5	7
	1500	1	1	1	3	4	6
	1750	1	1	1	2	4	5
	2000	0	1	1	1	3	5

Notes:

1. This table is for concentric stranded conductors only. For compact stranded conductors, Table C.12(A) should be used.

2. Two-hour fire-rated RHH cable has ceramifiable insulation which has much larger diameters than other RHH wires. Consult manufacturer's conduit fill tables.

*Types RHH, RHW, and RHW-2 without outer covering.

Table C.12(A) Maximum Number of Compact Conductors in Type EB, PVC Conduit
(Based on Table 1, Chapter 9)

		COMPACT CONDUCTORS					
	Conductor Size (AWG/ kcmil)	Metric Designator (Trade Size)					
Type		53 (2)	78 (3)	91 (3½)	103 (4)	129 (5)	155 (6)
THW,	8	30	68	89	113	174	247
THW-2,	6	23	52	69	87	134	191
THHW	4	17	39	51	65	100	143
	2	13	29	38	48	74	105
	1	9	20	26	34	52	74
	1/0	8	17	23	29	45	64
	2/0	6	15	19	24	38	54
	3/0	5	12	16	21	32	46
	4/0	4	10	14	17	27	38
	250	3	8	11	14	21	30
	300	3	7	9	12	19	26
	350	3	6	8	11	17	24
	400	2	6	7	10	15	21
	500	1	5	6	8	12	18
	600	1	4	5	6	10	14
	700	1	3	4	6	9	13
	750	1	3	4	5	8	12
	900	1	3	4	5	7	10
	1000	1	2	3	4	7	9
THHN,	8	—	—	—	—	—	—
THWN,	6	34	77	100	128	196	279
THWN-2	4	21	47	62	79	121	172
	2	15	34	44	57	87	124
	1	11	25	33	42	65	93
	1/0	9	22	28	36	56	79
	2/0	8	18	23	30	46	65
	3/0	6	15	20	25	38	55
	4/0	5	12	16	20	32	45
	250	4	10	13	16	25	35
	300	4	8	11	14	22	31
	350	3	7	9	12	19	27
	400	3	6	8	11	17	24
	500	2	5	7	9	14	20
	600	1	4	6	7	11	16
	700	1	4	5	6	10	14
	750	1	4	5	6	9	14
	900	1	3	4	5	7	10
	1000	1	3	3	4	7	10
XHHW,	8	39	88	115	146	225	320
XHHW-2	6	29	65	85	109	167	238
	4	21	47	62	79	121	172
	2	15	34	44	57	87	124
	1	11	25	33	42	65	93
	1/0	9	22	28	36	56	79
	2/0	8	18	24	30	47	67
	3/0	6	15	20	25	38	55
	4/0	5	12	16	21	32	46
	250	4	10	13	17	26	37
	300	4	8	11	14	22	31
	350	3	7	10	12	19	28
	400	3	7	9	11	17	25
	500	2	5	7	9	14	20
	600	1	4	6	7	11	16
	700	1	4	5	6	10	14
	750	1	3	5	6	9	13
	900	1	3	4	5	7	10
	1000	1	3	4	5	7	10

Definition: *Compact stranding* is the result of a manufacturing process where the standard conductor is compressed to the extent that the interstices (voids between strand wires) are virtually eliminated.

Chapter 6

How to Design a Lighting System

With the increased concern for energy conservation in recent years, much attention has been focused on lighting energy consumption and methods for reducing it. Along with this concern for energy efficient lighting has come the realization that lighting has profound effects on worker productivity as well as important aesthetic qualities. This chapter presents an introduction to lighting design and some of the energy efficient techniques that can be utilized while maintaining the quality of illumination.

LIGHTING EFFICIENCY

Lighting Basics

About 20 percent of all electricity generated in the United States is used for lighting. By understanding the basics of lighting design, several ways to improve the efficiency of lighting systems will become apparent.

There are two common lighting methods used. One is called the lumen method, while the other is the point-by-point method. The lumen method assumes an equal footcandle level throughout the area. This method is used frequently by lighting designers since it is simplest; however, it wastes energy, since it is the light "at the task" which must be maintained and not the light in the surrounding areas. The point-by-point method calculates the lighting requirements for the task in question.

Point-by-point Method

The point-by-point method makes use of the inverse square law, which states that the illuminance at a point on a surface perpendicular to the light ray is equal to the luminous intensity of the source at that point divided by the square of the distance between the source and the point of calculation, as illustrated in Formula 6-1.

$$E = \frac{I}{D^2} \qquad \text{Formula (6-1a)}$$

$$E = \frac{I}{D^2} \ (\text{Co}S\Theta) \qquad \text{Formula (6-1b)}$$

Where
    E = Illuminance in footcandles
    I = Luminous intensity in candles
    D = Distance in feet between the source and the point
        of calculation.
    Θ = Angle between the source and the point of calculation.

Lumen Method

    A footcandle is the illuminance on a surface of one square foot in area having a uniformly distributed flux of one lumen. From this definition, the lumen method is developed and illustrated by Formula 6-2.

$$N = \frac{F_1 \times A}{Lu \times L_1 \times L_2 \times L_3 \times Cu} \qquad \text{Formula (6-2)}$$

Where
    N is the number of lamps required.
    F_1 is the required footcandle level at the task. A footcandle is a measure of illumination—one standard candle power measured one foot away.
    A is the area of the room in square feet.
    Lu is the lumen output per lamp. A lumen is a measure of lamp intensity; its value is found in the manufacturer's catalogue.
    Cu is the coefficient of utilization. It represents the ratio of the lumens reaching the working plane to the total lumens generated by the lamp. The coefficient of utilization makes allowances for light absorbed or reflected by walls, ceilings, and the fixture itself. Its values are found in the manufacturer's catalogue.
    L_1 is the lamp depreciation factor. It takes into account that the lamp lumen depreciates with time. Its value is found in the manufacturer's catalogue.
    L_2 is the luminaire (fixture) dirt depreciation factor. It takes into account the effect of dirt on a luminaire and varies with type of luminaire and the atmosphere in which it is operated.
    L_3 is the lamp burn out factor

The lumen method formula illustrates several ways lighting efficiency can be improved.

Faced with the desire to reduce their energy use,* lighting consumers have four options: 1) reduce light levels, 2) purchase more efficient equipment, 3) provide light when needed at the task at the required level, and 4) add control and reduce lighting loads automatically. The multitude of equipment options to meet one or more of the above needs permits the consumer and the lighting designer-engineer to consider the trade-offs between the initial and operating costs, based upon product performance (life, efficacy, color, glare, and color rendering).

Some definitions and terms used in the field of lighting will be presented to help consumers evaluate and select lighting products best suiting their needs. Then, some state-of-the-art advances will be characterized so that their benefits and limitations are explicit.

Lighting Terminology

Efficacy is the amount of visible light (lumens) produced for the amount of power (watts) expended. It is a measure of the efficiency of a process, but it is a term used in place of efficiency when the input (W) has different units than the output (lm) and expressed in lm/W.

Color Temperature is a measure of the color of a light source relative to a black body at a particular temperature expressed in degrees Kelvin (°K). Incandescents have a low color temperature (~2800°K) and have a red-yellowish tone; daylight has a high color temperature (~6000°K) and appears bluish. Today, the phosphors used in fluorescent lamps can be blended to provide any desired color temperature in a range from 2800°K to 6000°K.

Color Rendering is a parameter that describes how a light source renders a set of colored surfaces with respect to a black body light source at the same color temperature. The color rendering index (CRI) runs from 0 to 100. It depends upon the specific wavelengths of which the light is composed. A black body has a continuous spectrum and contains all of the colors in the visible spectrum. Fluorescent lamps and high intensity discharge lamps (HID) have a spectrum rich in certain colors and devoid in others. For example, a light source that is rich in blues and low in reds could appear white, but when it is reflected from a substance, it would make red materials appear faded. The same material would appear

---

*Source: Lighting Systems Research, R.R. Verderber

different when viewed with an incandescent lamp, which has a spectrum that is rich in red.

LIGHT SOURCES*

Figure 6-1 indicates the general lamp efficacy ranges for the generic families of lamps most commonly used for both general and supplementary lighting systems. Each of these sources is discussed briefly here. It is important to realize that in the case of fluorescent and high intensity discharge lamps, the figures quoted for "lamp efficacy" are for the lamp only and do not include the associated ballast losses. To obtain the total system efficacy, ballast input watts must be used, rather than lamp watts, to obtain an overall system lumen per watt figure. This will be discussed in more detail in a later section.

Incandescent lamps have the lowest lamp efficacies of the commonly used lamps. This would lead to the accepted conclusion that incandescent lamps should generally not be used for large-area, general lighting systems where a more efficient source could serve satisfactorily. However, this does not mean that incandescent lamps should never be used. There are many applications where the size, convenience, ease of control, color rendering, and relatively low cost of incandescent lamps are suitable for a specific application.

General service incandescent lamps do not have good lumen maintenance throughout their lifetime. This is the result of the tungsten's evaporation off the filament during heating as it deposits on the bulb wall, thus darkening the bulb and reducing the lamp lumen output.

Efficient Types of Incandescents for Limited Use

Attempts to increase the efficiency of incandescent lighting while maintaining good color rendition have led to the manufacture of a number of energy-saving incandescent lamps for limited residential use.

Tungsten Halogen lamps vary from the standard incandescent by the addition of halogen gases to the bulb. Halogen gases keep the glass bulb from darkening by preventing the filament's evaporation, thereby increasing lifetime up to four times that of a standard bulb. The lumen-per-watt rating is approximately the same for both types of incandescents, but

---

*Source: Selection Criteria for Lighting Energy Management, Roger L. Knott

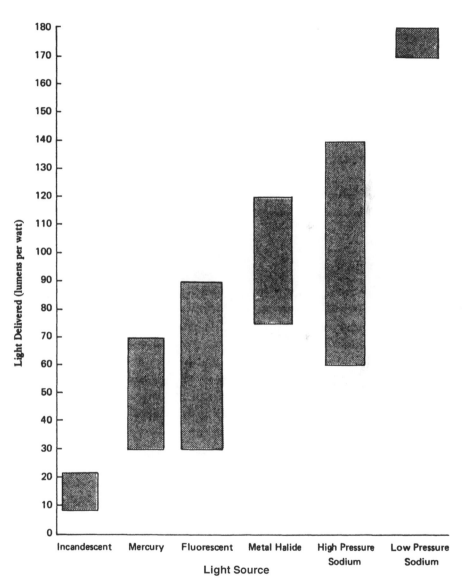

Figure 6-1. Efficacy of Various Light Sources

tungsten halogen lamps average 94% efficacy throughout their extended lifetime, offering significant energy and operating cost savings. Tungsten halogen lamps may require special fixtures, and during operation the surface of the bulb reaches very high temperatures. Encapsulated tungsten halogen sealed lamps are available that fit a standard incandescent base for use in the home. Sealed tungsten halogen lamps are also available for car headlamps.

R-Lamps are incandescents with an interior coating of aluminum that directs the light to the front of the bulb. Certain incandescent light fixtures, such as recessed or directional fixtures, trap light inside. Reflector lamps project a cone of light out of the fixture and into the room, so that more light is delivered where it is needed. In these fixtures, a 50-watt reflector bulb will provide better lighting and use less energy when substituted for a 100-watt standard incandescent bulb.

R-lamps are an appropriate choice for task lighting (because they directly illuminate a work area) and for accent lighting. Reflector lamps are available in 25, 30, 50, 75, and 150 watts. While they have a lower initial efficiency (lumens per watt) than regular incandescents, they direct light more effectively, so more light is actually delivered than with regular incandescents. (See Figure 6-2).

PAR Lamps (parabolic aluminized reflector lamps are reflector lamps) with a lens of heavy, durable glass that makes them an appropriate choice for outdoor flood and spot lighting. They are available in 75, 150, and 250 watts. They have longer lifetimes with less depreciation than standard incandescents.

ER Lamps (ellipsoidal reflector lamps) are ideally suited for recessed fixtures, because the beam of light produced is focused two inches ahead of the lamp to reduce the amount of light trapped in the fixture. In a directional fixture, a 75-watt ellipsoidal reflector lamp delivers more light than a 150-watt R-lamp. (See Figure 6-2.)

Mercury vapor lamps find very limited use in today's lighting systems, because fluorescent and other high intensity discharge (HID) sources have surpassed them in both lamp efficacy and system efficiency. Typical ratings for mercury vapor lamps range from about 30 to 70 lumens per watt. Fluorescent systems are available today that can do many of the jobs mercury used to do, and they do it more efficiently. Mercury vapor lamps are not common, as they have been replaced by fluorescent and higher efficacy HID sources.

Fluorescent lamps have made dramatic advances. Several styles of low

Standard Incandescent

R-Lamp

A high percentage
of light output
is trapped in fixture

An aluminum
coating directs light
out of the fixture

ER Lamp

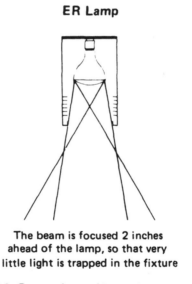

The beam is focused 2 inches
ahead of the lamp, so that very
little light is trapped in the fixture

Figure 6-2. Comparison of Incandescent Lamps

wattage, compact lamps have provided a steady parade of products. Lamp efficacy now ranges from about 30 lumens per watt to near 90 lumens per watt, or more. A range of colors is available, and lamp manufacturers have produced fluorescent and metal halide lamps that have much more consistent color-rendering properties, allowing greater flexibility in mixing of fluorescent and metal halide sources without creating disturbing color mismatches. The compact fluorescent lamps (plug-in and screw-in types) permit design of much smaller luminaries.

Metal halide lamps fall into a lamp efficacy range of approximately 75-125 lumens per watt. This makes them somewhat less energy efficient than high pressure sodium. Metal halide lamps generally have fairly good color rendering qualities. While this lamp displays some very desirable qualities, it also has some distinct drawbacks, including relatively shorter life for an HID lamp and long restrike time to restart after the lamp has been shut off. This source is used very successfully in many applications, especially where color is critical.

High pressure sodium lamps have extremely high efficacy (60-140 lumens/watt) in a lamp that operates in fixtures having construction very similar to those used for metal halide. High quality lamps and ballasts provide very satisfactory service. The 24,000-hour lamp life, good lumen maintenance, and high efficacy of these lamps make them ideal sources for industrial and outdoor applications where discrimination of a range of colors is not critical.

The lamp's primary drawback is the rendering of some colors. The lamp produces a high percentage of light in the yellow range of the spectrum. This tends to accentuate colors in the yellow region. Rendering of reds and greens shows a pronounced color shift. This can be compensated for in the selection of the finishes for the surrounding areas, and if properly done, the results can be very pleasing. In areas where color selection, matching and discrimination are necessary, high pressure sodium should not be used as the only source of light. It is possible to gain quite satisfactory color rendering by mixing high pressure sodium and metal halide in the proper proportions. Since both sources have relatively high efficacies, there is not a significant loss in energy efficiency by making this compromise.

High pressure sodium has been used quite extensively in outdoor applications for roadway, parking and façade or security lighting. This source will yield a high efficiency system; however, it should be used only with the knowledge that foliage and landscaping colors will be severely distorted where high pressure sodium is the only, or predominant, illuminant. Used as a parking lot source, there may be some difficulty in identification of vehicle colors in the lot. If this is a problem, it is necessary for the designer or owner to determine the extent and significance of this problem and what steps might be taken to alleviate it.

High pressure sodium lamps with improved color rendering qualities are available; however, the efficacy of the color-improved lamps is somewhat lower.

Low pressure sodium lamps provide the highest efficacy of any of the sources for general lighting, with values ranging up to 180 lumens per watt. Low pressure sodium lamps are practically monochromatic and produce an almost pure yellow light with very high efficacy, rendering all colors gray except yellow or near yellow. This effect results in no color discrimination under low pressure sodium lighting and is suitable for use in a very limited number of applications. It is an acceptable source for warehouse lighting where it is only necessary to read labels but not to choose items by color. This source has application for either indoor/ outdoor safety or security lighting, as long as color rendering is not important.

Xenon lamps use ionized xenon gas. The xenon lamps produce a very white light close to a full spectrum light. The xenon lamp has a glass or quartz tube with tungsten electrodes at end. The use of xenon lamps in car headlights have expanded the lamp's popularity.

Light Emitting Diodes (LEDs) are being used in a variety of ways for lighting and are obtaining brighter outputs. A major example of LED's use is for the traffic and other signaling lamps. These electronic lights are very efficient and have a very long life. LEDs often use 15% or less energy than typical halogen incandescent lamps. Due to the low energy requirements of LED's, they can often be used with solar panels, especially in remote areas.

Quartz Halogen Lamps—Use quartz crystal encapsulation rather than boron silica glass. The quartz encapsulation is filled with halogen gas. The quartz filled with the halogen gas allows higher burning temperature and more light out per watts in.

Electronics ballasts reduce lamp-blackening, improve color consistency, and practically eliminate flicker, for an overall improvement in light quality. They are more efficacious than other ballasts.

Programmed start ballasts are used for energy savings, using automated light controls to switch lights off when an area is not used or filled with external illumination.

An *HID electronic ballast* system features better lumen maintenance and lower ballast losses, compared to magnetic HID ballasts. The HID electronics ballasts reduce energy consumption and have lower maintenance requirements. As a result, fewer fixtures are required due to lower lamp loss factors in new installations.

In addition to these primary sources, there are a number of retrofit lamps which allow use of higher efficacy sources in the sockets of existing fixtures. Therefore, metal halide or high pressure sodium lamps can be retrofitted into mercury vapor fixtures, or self-ballasted mercury lamps can replace incandescent lamps. These lamps all make some compromises in operating characteristics, life and/or efficacy.

Figure 6-3 presents data on the efficacy of each of the major lamp types in relation to the wattage rating of the lamps. Without exception, the efficacy of the lamp increases as the lamp wattage rating increases.

The lamp efficacies discussed here have been based on the lumen output of a new lamp after 100 hours of operation or the "initial lumens." Not all lamps age in the same way. Some lamp types, such as lightly loaded fluorescent and high pressure sodium, hold up well and maintain their lumen output at a relatively high level until they are into or past middle age. Others, as represented by heavily loaded fluorescent, mercury vapor, and metal halide decay rapidly during their early years and then coast along at a relatively lower lumen output throughout most of their

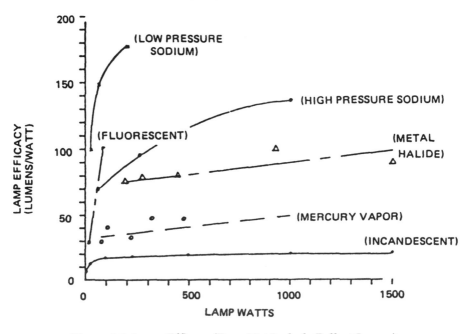

Figure 6-3. Lamp Efficacy (Does Not Include Ballast Losses)

useful life. These factors must be considered when evaluating the various sources for overall energy efficacy.

Incandescent Replacement

The most efficacious lamps that can be used in incandescent sockets are compact fluorescent lamps. The most popular systems are the twin tubes and double-twin tubes. These are closer to the size and weight of the incandescent lamp than the earlier type of fluorescent (circline) replacements.

Twin tubes with lamp wattages from 5 to 13 watts provide amounts of light ranging from 240 to 850 lumens. Table 6-1 lists the characteristics of various types of incandescent and compact fluorescent lamps that can be used in the same type sockets.

Table 6-1. Lamp Characteristics

Lamp Type (Total Input Power) *	Lamp Power (W)	Light Output (lumens)	Lamp Life (hour)	Efficacy (lm/W)
100 W (Incandescent)	100	1750	750	18
75 W (Incandescent)	75	1200	750	16
60 W (Incandescent)	60	890	1000	15
40 W (Incandescent)	40	480	1500	12
25 W (Incandescent)	25	238	2500	10
22 W (Fl. Circline)	18	870	9000	40
44 W (Fl. Circline)	36	1750	9000	40
7 W (Twin)	5	240	10000	34
10 W (Twin)	7	370	10000	38
13 W (Twin)	9	560	10000	43
19 W (Twin)	13	850	10000	45
18 W (Solid-State)**	—	1100	7500	61

*Includes ballast losses.
**Operated at high frequency.

The advantages of the compact fluorescent lamps are larger and increased efficacy, longer life, and reduced total cost. The cost per 1000 hours hours of operating the 100-watt incandescent and the 44-watt fluorescent is $15.00 and $6.67, respectively. This is based upon an energy cost of $.15 per kWh. The circline lamps were much larger and heavier than the incandescents and would fit in a limited number of fixtures. The twin tubes are only slightly heavier and larger than the equivalent incandescent lamp. However, there are some fixtures that are too small for them to be employed.

The narrow tube diameter compact fluorescent lamps are now possible because of rare earth phosphors. These phosphors improve lumen depreciation at high lamp power loadings. The second important characteristic of these narrow band phosphors is their high efficiency in converting the ultraviolet light generated in the plasma into visible light. By proper mixing of these phosphors, the color characteristics (color temperature and color rendering) are similar to the incandescent lamp.

Two types of compact fluorescent lamps are screw-in and plug-in. In one type of lamp system, the ballast and lamp are integrated into a single package. In the second type, the lamp and ballast are separate, and when a lamp burns out it can be replaced. In the integrated system, both the lamp and the ballast are discarded when the lamp bums out.

It is important to recognize when purchasing these compact fluorescent lamps that they provide the equivalent light output of the lamps being replaced. The initial lumen output for the various lamps is shown in Table 6-1.

Lighting Efficiency Options

Several lighting efficiency options are illustrated below (refer to Formula 6-1).

Footcandle Level—The footcandle level required is that level required at the task. For example offices require 100 fc. Footcandle levels can be lowered to one third of the levels for surrounding areas such as aisles. (A minimum 20-footcandle level should be maintained.)

The placement of the lamp is also important. If the luminaire can be lowered or placed at a better location, the lamp wattage for the required lighting level may be reduced .

Coefficient of Utilization (Cu)—The color of the walls, ceiling, and floors, the type of luminaire, and the characteristics of the room determine the Cu. This value is determined based on manufacturer's literature. The Cu can be improved by analyzing components such as lighter colored walls and more efficient luminaires for the space.

Lamp Depreciation Factor and Dirt Depreciation Factor—These two factors are involved in the maintenance program. Choosing a luminaire which resists dirt buildup, group relamping, and cleaning the luminaire will keep the system at optimum performance. Taking these factors into account can reduce the number of lamps initially required.

The *Light Loss Factor* (*LLF*) takes into account that the lamp lumen depreciates with time (L_1), that the lumen output depreciates due to dirt buildup (L_2), and that lamps burn out (L_3). Formula 6-3 illustrates the relationship of these factors.

$$LLF = L_1 \times L_2 \times L_3 \qquad \text{Formula (6-3)}$$

To reduce the number of lamps required, which in turn reduces energy consumption, it is necessary to increase the overall light loss factor. This is accomplished in several ways. One is to choose the luminaire which minimizes dust buildup. The second is to improve the maintenance program to replace lamps prior to burn-out. Thus, if it is known that a group relamping program will be used at a given percentage of rated life, the appropriate lumen depreciation factor can be found from manufacturer's data. It may be decided to use a shorter relamping period in order to increase (L_1) even further. If a group relamping program is used, (L_3) is assumed to be unity.

Figure 6-4 illustrates the effect of dirt buildup on (L_2) on a dustproof luminaire. Every luminaire has a tendency for dirt buildup. Manufacturer's data should be consulted when estimating (L_2) for the luminaire in question.

Solid-state Ballasts

Fluorescent lamps at high frequency (20 to 30 kHz) with solid-state ballasts have achieved credibility. The fact that all of the major ballast manufacturers offer solid-state ballasts and that the major lamp companies

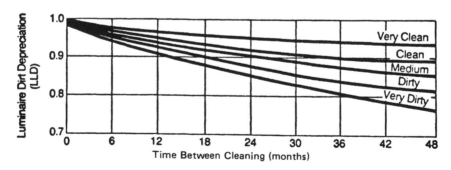

Figure 6-4. Effect of Dirt Buildup on Dustproof Luminaires for Various Atmospheric Conditions

have designed new lamps to be operated at high frequency is evidence that the solid-state high frequency ballast is now state-of-the-art.

It has been shown that fluorescent lamps operated at high frequency are 10 to 15 percent more efficacious than 60 Hz operation. In addition, the solid-state ballast is more efficacious than conventional ballasts in conditioning the input power for the lamps, such that the total system efficacy increase is between 20 and 25 percent. That is, for a standard two-lamp, 40-watt rapid-start system, overall efficacy is increased from 63 lm/W to over 80 lm/W.

Continued development of the product has improved reliability and reduced cost. In order to be more competitive with initial costs, there are three- and four-lamp ballasts for lamps. These multi-lamp ballasts reduce the initial cost per lamp, as well as the installation cost, and are even more efficient than the one- and two-lamp ballast system.

The American National Standards Institute (ANSI) ballast committee has standards for solid-state ballasts. The ballast factor is one parameter that will be specified by ANSI. However, solid-state ballasts are available with different ballast factors. The ballast factor is the light output provided by the ballast-lamp system compared to the light output of the lamp specified by the lamp manufacturer. The ANSI ballast factor standard for 40-watt F40 fluorescent lamps is 95 ± 2.5 percent.

There are solid-state ballasts with a ballast factor exceeding 100 percent. These ballasts are most effectively used in new installations. In these layouts, more light from each luminaire will reduce the number of luminaires, ballasts and lamps, hence reducing both initial and operating costs. It is essential that the lighting designer/engineer and consumer know the ballast factor for the lamp-ballast system. The ballast factor for a ballast also depends upon the lamp. For example, a core-coil ballast will have a ballast factor of 95 ± 2 percent when operating a 40-watt F40 argon-filled lamp, and less when operating an "energy saving" 34-watt F40 Krypton-filled lamp. The ballast factor instead will be about 87 ± 2.5 percent with the 34-watt energy saving lamp.

Table 6-2 compares several types of solid-state ballasts with a standard core-coil ballast that meets the ANSI standard with the two-lamp, 40-watt F40 lamp. Notice that the system efficacy of any ballast system is about the same operating a 40-watt or a 34-watt F40 lamp. Although the 34-watt "lite white" lamp is about 6 percent more efficient than the 40-watt "cool white" lamp, the ballast losses are greater with the 34-watt lamp, due to an increased lamp current. The lite white phosphor is more efficient than the cool white phosphor but has poorer color rendering characteristics.

Note that the percent flicker is drastically reduced when the lamps are operated at high frequency with solid-state ballasts. A scientific field study of office workers in the U.K. showed that complaints of headaches and eye-strain are 50 percent less under high frequency lighting when compared to lamps operating at 50 cycles, the line frequency in the U.K.

Table 6-2. Performance of F40 Fluorescent Lamp Systems

Characteristic	Core-Coil		—Solid-state Ballasts —				
	2 Lamps, T-12		2 Lamps, T-12		4 Lamps, T-12		2 Lamps, T-8
	40W	34W	40W	34W	40W	34W	32W
Power (W)	96	79	72	63	136	111	65
Power Factor (%)	98	92	95	93	94	94	89
Filament Voltage (V)	3.5	3.6	3.1	3.1	2.0	1.6	0
Light Output (lm)	6050	5060	5870	5060	11,110	9250	5820
Ballast Factor	.968	.880	.932	.865	.882	.791	1.003
Flicker (%)	30	21	15	9	1	0	1
System Efficacy (lm/W)	63	64	81	81	82	83	90

Each of the above ballasts has different factors, which are lower when operating the 34-watt Krypton-filled lamp. Table 6-2 also lists the highest system efficacy of 90 lumens per watt for the solid-state ballast and T-8, 32-watt lamp.

All of the above solid-state ballasts can be used in place of core-coil ballasts specified to operate the same lamps. To determine the illumination levels, or the change in illumination levels, the manufacturer must supply the ballast factor for the lamp type employed. The varied light output from the various systems allows the lighting designer/engineer to precisely tailor the lighting level.

Table 6-2 compares the cost of lighting for the two-lamp, F40 system and the two types of ballasts, and for a ballast operating the 40-watt and 34-watt lamps. Included in the table is the T-8, 32-watt lamp system, which has an efficacy of 90 lumens per watt. The table shows that the cost of illumination per million lumen hours is least for the most efficient solid-state ballast system and slightly less for all ballasts operating the 40-watt argon-filled lamps. The table uses realistic average costs for the ballasts and lamps. Consumers should use quoted prices as an element in their decision process, since prices will depend upon the volume purchased. The cost of fluorescent systems is much less than incandescent lamps systems.

CONTROL EQUIPMENT

Table 6-3 lists various types of equipment that can be components of a lighting control system, with a description of the predominant characteristic of each type of equipment. Static equipment can alter light levels semipermanently. Dynamic equipment can alter light levels automatically over short intervals to correspond to the activities in a space. Different sets of components can be used to form various lighting control systems in order to accomplish different combinations of control strategies.

Table 6-3. Lighting Control Equipment

System	Remarks
STATIC:	
Delamping	Method for reducing light level 50%.
Impedance Monitors	Method for reducing light level 30, 50%.
DYNAMIC:	
Light Controllers	
Switches/Relays	Method for on-off switching of large banks of lamps.
Voltage/Phase Control	Method for controlling light level continuously 100 to 50%.
Solid-State Dimming Ballasts	Ballasts that operate fluorescent lamps efficiently and can dim them continuously (100 to 10%) with low voltage.
SENSORS:	
Clocks	System to regulate the illumination distribution as a function of time.
Personnel	Sensor that detects whether a space is occupied by sensing the motion of an occupant.
Photocell	Sensor that measures the illumination level of a designated area.
COMMUNICATION:	
Computer/Microprocessor	Method for automatically communicating instructions and/or input from sensors to commands to the light controllers.
Power-Line Carrier	Method for carrying information over existing power lines rather than dedicated hard-wired communication lines.

FLUORESCENT LIGHTING CONTROL SYSTEMS

The control of fluorescent lighting systems is receiving increased attention. Two major categories of lighting control are available personnel sensors and lighting compensators.

Personnel Sensors

There are three classifications of personnel sensors-ultrasonic, infrared and audio.

Ultrasonic sensors generate sound waves outside the human hearing range and monitor the return signals. Ultrasonic sensor systems are generally made up of a main sensor unit with a network of satellite sensors providing coverage throughout the lighted area. Coverage per sensor is dependent upon the sensor type and ranges between 500 and 2,000 square feet. Sensors may be mounted above the ceiling, suspended below the ceiling, or mounted on the wall. Energy savings are dependent upon the room size and occupancy. Advertised savings range from 20 to 40 percent.

Infrared sensor systems consist of a sensor and control unit. Coverage is limited to approximately 130 square feet per sensor. Sensors are mounted on the ceiling and usually directed towards specific work stations. They can be tied into the HVAC control and limit its operation also. Advertised savings range between 30 and 50 percent.

Audio sensors monitor sound within a working area. The coverage of the sensor is dependent upon the room shape and the mounting height. Some models advertise coverage of up to 1,600 square feet. The first cost of the audio sensors is approximately one half that of the ultrasonic sensors. Advertised energy savings are approximately the same as the ultrasonic sensors. Several restrictions apply to the use of the audio sensors. First, normal background noise must be less than 60 dB. Second, the building should be at least 100 feet from the street and may not have a metal roof.

Lighting Compensators

Lighting compensators are divided into two major groups switched and sensored.

Switched compensators control the light level using a manually operated wall switch. These particular systems are used frequently in residential settings and are commonly known as "dimmer switches." Based on discussions with manufacturers, the switched controls are available for the 40-watt standard fluorescent bulbs only. The estimated savings are difficult to determine, as usually switched control systems are used to control room mood. The only restriction to their use is that the luminaire must have a dimming ballast.

Sensored compensators are available in three types. They may be very simple or very complex. They may be integrated with the building's energy management system or installed as a stand-alone system. The first type of system is the excess light turn-off (ELTO) system. This system senses daylight levels and automatically turns off lights as the sensed light level approaches a programmed upper limit. Advertised paybacks for these types of systems range from 1.8 to 3.8 years.

The second type of system is the daylight compensator (DAC) system. This system senses daylight levels and automatically dims lights to achieve a programmed room light level. Advertised savings range from 40 to 50 percent. The primary advantage of this system is that it maintains a uniform light level across the controlled system area. The third system type is the daylight compensator + excess light turn-off system. As implied by the name, this system is a combination of the first two systems. It automatically dims light outputs to achieve a designated light level and, as necessary, automatically turns off lights to maintain the desired room conditions.

Specular reflectors: Fluorescent fixtures can be made more efficient by the insertion of a suitably shaped specular reflector. The specular reflector material types are aluminum, silver and multiple dielectric film mirrors. The latter two have the highest reflectivity, while the aluminum reflectors are less expensive.

Measurements show the fixture efficiency with higher reflectance specular reflectors (silver or dielectric films) is improved by 15 percent compared to a new fixture with standard diffuse reflectors.

Specular reflectors tend to concentrate more light downward, with reduced light at high exit angles. This increases the light modulation in the space, which is the reason several light readings at different sites around the fixture are required for determining the average illuminance. The increased downward component of candle power may increase the potential for reflected glare from horizontal surfaces.

When considering reflectors, information should be obtained on the new candle power characteristics. With this information a lighting designer or engineer can estimate the potential changes in modulation and reflected glare.

LIGHTING DESIGN

In the previous section, it was seen that a lamp produces an amount of light (measured in lumens) that depends on the power consumed and the type of lamp. Equally important to the amount of light produced by a lamp is the amount of light which is "usable," or provides illumination for the desired task. Luminaires, or lighting fixtures, are used to direct the light to a usable location, depending on the specific requirements of the area to be lighted. Regardless of the luminaire type, some of the light is directed in non-usable directions, is absorbed by the luminaire itself, or is absorbed by the walls, ceiling or floor of the room.

The coefficient of utilization, or CU, is a factor used to determine the efficiency of a fixture in delivering light for a specific application. The coefficient of utilization is determined as a ratio of light output from the luminaire that reaches the workplane to the light output of the lamps alone. Luminaire manufacturers provide CU data in their catalogs which are dependent on room size and shape, fixture mounting height, and surface reflectances. Table 6-4 illustrates the form in which a vendor summarized the data used for determining the coefficient of utilization.

To determine the coefficient of utilization, the room cavity ratio, wall reflectance, and effective ceiling cavity reflectance must be known.

Most data assume a 20% effective floor cavity reflectance. To determine the coefficient of utilization, the following steps are needed:
   (a) Estimate wall and ceiling reflectances.
   (b) Determine room cavity ratio.
   (c) Determine effective ceiling reflectance (pCC).

Step (a)
Typical reflectance values are shown in Table 6-5.
Steps (b) and (c)
Once the wall and ceiling reflectances are estimated, it is necessary to analyze the room configuration to determine the effective reflectances. Any room is made up of a series of cavities that have effective reflectances with respect to each other and the work plane. Figure 6-5 indicates the basic cavities.

The space between fixture and ceiling is the ceiling cavity. The space between the work plane and the floor is the floor cavity. The space between the fixture and the work plane is the room cavity. To determine the cavity ratio, use Figure 6-5 to define the cavity depth, and then use Formula 6-4.

Table 6-4. Typical Lamp Vendor Data

Coefficients of Utilization/Effective Floor Cavity Reflectance 20% (pFC)

% REFLECTANCE EFF. CEIL. (pCC)	WALL (pW)	ROOM CAVITY RATIO									
		1	2	3	4	5	6	7	8	9	10
80	50	0.854	0.779	0.711	0.647	0.591	0.539	0.490	0.446	0.407	0.355
	30	0.828	0.739	0.664	0.594	0.533	0.481	0.432	0.388	0.349	0.296
	10	0.805	0.705	0.626	0.552	0.491	0.440	0.392	0.347	0.309	0.258
70	50	0.832	0.761	0.698	0.635	0.578	0.530	0.483	0.438	0.401	0.349
	30	0.808	0.724	0.653	0.585	0.526	0.475	0.426	0.384	0.345	0.295
	10	0.786	0.695	0.618	0.546	0.486	0.434	0.387	0.344	0.308	0.256
50	50	0.788	0.725	0.669	0.610	0.558	0.511	0.466	0.424	0.388	0.339
	30	0.770	0.696	0.632	0.568	0.513	0.464	0.416	0.375	0.338	0.288
	10	0.754	0.670	0.602	0.534	0.478	0.428	0.382	0.339	0.303	0.253
30	50	0.750	0.694	0.642	0.587	0.539	0.495	0.450	0.412	0.377	0.329
	30	0.736	0.671	0.612	0.552	0.499	0.453	0.408	0.367	0.331	0.282
	10	0.722	0.649	0.586	0.523	0.469	0.421	0.375	0.335	0.299	0.249
10	50	0.716	0.665	0.618	0.566	0.521	0.479	0.438	0.399	0.366	0.319
	30	0.704	0.645	0.592	0.536	0.487	0.442	0.399	0.360	0.325	0.276
	10	0.693	0.628	0.571	0.511	0.460	0.413	0.370	0.330	0.294	0.245

Table 6-5. Typical Reflection Factors

COLOR	REFLECTION FACTOR
White and very light tints	.75
Medium blue, green, yellow or gray	.50
Dark gray, medium blue	.30
Dark blue, brown, dark green, and wood finishes	.10

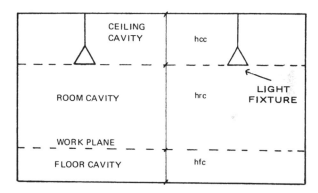

Figure 6-5. Cavity Configurations

$$\text{Cavity Ratio} = \frac{5 \times d \times (L+W)}{L \times W} \qquad \text{(Formula 6-4)}$$

Where
    d = Depth of the cavity as defined in Figure 6-5
    L = Room (or area) length
    W = Room (or area) width

To determine the effective ceiling or floor cavity reflectance, proceed in the same manner to define the ceiling or floor cavity ratio, then refer to Table 6-4 to find the corresponding effective ceiling or floor cavity reflectance.

SIM 6-1

For Process Plant No.1, determine the coefficient of utilization for a room which measures 24′ × 100′. The ceiling is 20′ high and the fixture is mounted 4′ from the ceiling. The tasks in the room are performed on work benches 3′ above the floor. Use the data in Table 6-4.

Answer

Step (a)

Since no wall or ceiling reflectance data were given, assume a ceiling of .70 and wall of .5.

Step (b)

Assume 3' working height.

hrc = 20-4-3 = 13 (from Figure 6-5)

From Formula 6-4, RCR = 3.4

Step (c)

From Figure 6-1, hcc = 4

From Formula 6-4, CCR = 1

From Table 6-6 (pp 190-191), pCC = 58

From Table 6-4, Coefficient of Utilization = 0.64 (interpolated)

FIXTURE LAYOUT

The fixture layout is dependent on the area. The initial layout should have equal spacing between lamps, rows and columns. The end fixture should be located at one half the distance between fixtures. The maximum distance between fixtures usually should not exceed the mounting height unless the manufacturer specifies otherwise. Figure 6-6 illustrates a typical layout. If the fixture is fluorescent, it may be more practical to run the fixtures together. Since the fixtures are 4 feet or 8 feet long, a continuous wireway will be formed.

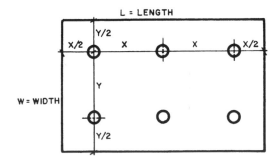

Figure 6-6. Typical Fixture Layout

SIM 6.2

Referring to SIM 6-1, determine the required number of fixtures and lighting layout to give an average, maintained footcandle level of 50. The light loss factor is estimated to be 0.7. The initial lumen output is 8,500 per lamp.

Answer

$$\text{No. of fixtures} = \frac{\text{Area} \times \text{Desired Maintained Footcandle}}{\text{Lumens} \times \text{CU} \times \text{LU}}$$

$$= \frac{24 \times 100 \times 50}{8{,}500 \times 0.64 \times 0.7} = \frac{31.5 \text{ or } 32}{\text{fixtures}}$$

		Rows	Columns	X Spacing	Y Spacing
Typical Combinations	(a)	4	8	12.5	6
	(b)	3	11	9	8
	(c)	2	16	6	12

(a)		(b)		(c)	
$8X = 100$		$11X = 100$		$16X = 100$	
$X = 12.5$		$X = 9$		$X = 6.2$	
$4Y = 24$		$3Y = 24$		$2Y = 24$	
$Y = 6$		$Y = 8$		$Y = 12$	

Alternate (b) is recommended, even though it requires one more fixture. It results in a good layout, illustrated as follows.

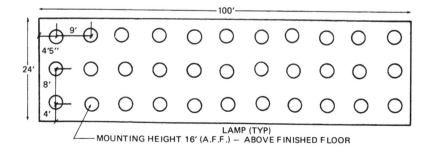

LAMP (TYP)
MOUNTING HEIGHT 16' (A.F.F.) – ABOVE FINISHED FLOOR

Table 6-6. Percent Effective Ceiling or Floor Cavity Reflectance for Various Reflectance Combinations

% Wall Reflectance → / Cavity Ratio ↓	90				80				70			50			30				10		
% Wall Reflectance	90	70	50	30	80	70	50	30	70	50	30	70	50	30	65	50	30	10	50	30	10
0	90	90	90	90	80	80	80	80	70	70	70	50	50	50	30	30	30	30	10	10	10
0.1	90	89	88	87	79	79	78	78	69	69	68	49	49	48	30	30	29	29	10	10	10
0.2	89	88	86	85	79	78	77	76	68	67	66	49	48	47	30	29	29	28	10	10	9
0.3	89	87	85	83	78	77	75	74	68	66	64	49	47	46	30	29	28	27	10	10	9
0.4	88	86	83	81	78	76	74	72	67	65	63	48	46	45	30	29	27	26	11	10	9
0.5	88	85	81	78	77	75	73	70	66	64	61	48	46	44	29	28	27	25	11	10	9
0.6	88	84	80	76	77	75	71	68	65	62	59	47	45	43	29	28	26	25	11	10	9
0.7	88	83	78	74	76	74	70	66	65	61	58	47	44	42	29	28	26	24	11	10	8
0.8	87	82	77	73	75	73	69	65	64	60	56	47	43	41	29	27	25	23	11	10	8
0.9	87	81	76	71	75	72	68	63	63	59	55	46	43	40	29	27	25	22	11	9	8
1.0	86	80	74	69	74	71	66	61	63	58	53	46	42	39	29	27	24	22	11	9	8
1.1	86	79	73	67	74	71	65	60	62	57	52	46	41	38	29	26	24	21	11	9	8
1.2	86	78	72	65	73	70	64	58	61	56	50	45	41	37	29	26	23	20	12	9	7
1.3	85	78	70	64	73	69	63	57	61	55	49	45	40	36	29	26	23	20	12	9	7
1.4	85	77	69	62	72	68	62	55	60	54	48	45	40	35	28	26	22	19	12	9	7
1.5	85	76	68	61	72	68	61	54	59	53	47	44	39	34	28	25	22	18	12	9	7
1.6	85	75	66	59	71	67	60	53	59	52	45	44	39	33	28	25	21	18	12	9	7
1.7	84	74	65	58	71	66	59	52	58	51	44	44	38	32	28	25	21	17	12	9	7
1.8	84	73	64	56	70	65	58	50	57	50	43	43	37	32	28	25	21	17	12	9	6
1.9	84	73	63	55	70	65	57	49	57	50	42	43	37	31	28	25	20	16	12	9	6
2.0	83	72	62	53	69	64	56	48	56	48	41	43	37	30	28	24	20	16	12	9	6

% Ceiling or Floor Reflectance (top header) — Ceiling or Floor Cavity Ratio (left axis)

(more)

2.1	2.2	2.3	2.4	2.5	2.6	2.7	2.8	2.9	3.0	3.1	3.2	3.3	3.4	3.5	3.6	3.7	3.8	3.9	4.0	4.1	4.2	4.3	4.4	4.5	4.6	4.7	4.8	4.9	5.0
6	6	6	6	6	5	5	5	5	5	5	5	5	5	5	5	4	4	4	4	4	4	4	4	4	4	4	4	4	4
9	9	9	9	9	9	9	9	9	8	8	8	8	8	8	8	8	8	8	8	8	8	8	8	8	8	8	8	8	8
13	13	13	13	13	13	13	13	13	13	13	13	13	13	13	13	13	13	13	13	13	13	13	13	14	14	14	14	14	14
16	15	15	14	14	13	13	13	12	12	12	11	11	11	11	10	10	10	10	9	9	9	9	8	8	8	8	8	7	7
20	19	19	19	18	18	18	18	17	17	17	16	16	16	16	15	15	15	15	15	14	14	14	14	14	14	13	13	13	13
24	24	24	24	23	23	23	23	23	22	22	22	22	22	22	21	21	21	21	21	21	20	20	20	20	20	20	19	19	19
28	28	28	28	27	27	27	27	27	27	27	27	27	27	26	26	26	26	26	26	26	26	26	26	25	25	25	25	25	25
29	29	28	27	27	26	26	25	25	24	24	23	23	22	22	21	21	21	20	20	20	19	19	19	19	18	18	18	18	17
36	36	35	35	34	34	33	33	33	32	32	31	31	31	30	30	30	29	29	29	28	28	28	27	27	27	26	26	26	26
43	42	42	42	41	41	41	41	40	40	40	40	39	39	39	39	38	38	38	38	37	37	37	37	37	36	36	36	36	36
40	39	38	37	36	35	34	33	33	32	31	30	30	29	29	28	27	27	26	26	25	25	25	24	24	24	23	23	23	22
47	46	46	45	44	43	43	42	41	40	40	39	39	38	38	37	37	37	36	35	35	34	34	34	33	33	33	32	32	32
56	55	54	54	53	53	52	52	51	51	50	50	49	49	48	48	48	47	47	46	46	46	45	45	45	44	44	44	44	43
47	45	44	43	42	41	40	39	38	38	37	36	35	34	33	33	32	31	30	30	29	29	28	28	27	26	26	25	25	25
55	54	53	52	51	50	49	48	48	47	46	45	44	44	43	42	42	41	40	40	39	39	38	38	37	37	36	36	35	35
63	63	62	61	61	60	60	59	58	58	57	57	56	56	55	54	54	53	53	52	52	51	51	51	50	50	49	49	49	48
69	68	68	67	67	66	66	66	65	65	64	64	64	63	63	62	62	62	61	61	60	60	60	59	59	59	58	58	58	57
52	51	50	48	47	46	45	44	43	42	41	40	39	38	37	36	35	35	34	33	32	32	31	30	30	29	29	28	28	27
61	60	59	58	57	56	55	54	53	52	51	50	49	48	48	47	46	45	45	44	43	43	42	41	41	40	40	39	38	38
71	70	69	68	68	67	66	66	65	64	64	63	62	62	61	60	60	59	59	58	57	57	56	56	55	55	54	54	53	53
83	83	83	82	82	82	82	81	81	81	80	80	80	80	79	79	79	79	78	78	78	78	78	77	77	77	77	76	76	76

Ceiling or Floor Cavity Ratio

CIRCUITING

Number of Lamps Per Circuit

A commonly used circuit loading is 1600 watts per lighting circuit breaker. This load includes fixture voltage and ballast loss. In SIM 6-2, assuming a ballast loss of 25 watts per fixture, 175 watt lamp, a 20 amp circuit breaker, and No. 12 gauge wire, eight lamps could be fed from each circuit breaker. A single-phase circuit panel is illustrated in Figure 6-7. (Note: In practice, ballast loss should be based on manufacturer's specifications.) Using these assumptions, solve problems SIM 6-3 and 6-4.

Room Size (in ft.)		Luminaires Lengthwise		Luminaires Crosswise	
		Ceiling Height (in feet)			
W	L	8.5	10.0	8.5	10.0
20 x	20	75	72	73	70
	30	75	72	73	70
	40	75	73	73	71
	60	75	73	72	71
30 x	20	78	75	77	73
	30	78	75	76	73
	40	77	75	75	72
	60	76	74	74	72
	80	76	74	73	72
40 x	20	81	78	80	77
	30	79	77	78	76
	40	78	77	76	75
	60	77	76	75	74
	80	77	75	74	73
	100	76	75	74	73
60 x	30	81	79	80	77
	40	79	78	78	76
	60	78	77	76	74
	80	77	76	75	74
	100	77	76	75	73
100 x	40	81	80	80	78
	60	80	78	78	76
	80	78	77	77	75
	100	78	76	76	74

**Wall Reflectance, 50%
Ceiling Cavity Reflectance, 80%
Floor Cavity Reflectance, 20%
Work Plane Illumination, 100fc

Figure 6-7. Single-Phase Circuit Panel

SIM 6-3

Next to each lamp place the panel designation and circuit number from which each lamp is fed; i.e., A-1, A-2, etc:.

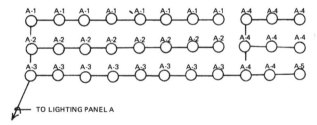

Answer

SIM 6-4

Designate a hot line from the circuit breaker with a small stroke and use a long stroke as a neutral; i.e., ─╫╫─ 4 wires, 2 hot and 2 neutrals. The lamps are connected with conduit as shown below. Designate the hot and neutrals in each branch.

Hint—start wiring from the last fixture in the circuit.

Answer

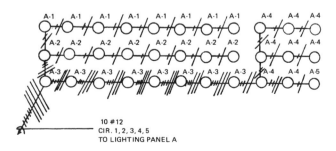

10 #12
CIR. 1, 2, 3, 4, 5
TO LIGHTING PANEL A

Note that only a single neutral wire is required for each 3 different phase wires.

POINTS ON LIGHTING DESIGN

- Identify all symbols for lighting fixtures.
- Include circuit numbers on all lights.
- Include a note on fixture mounting height.
- Show "homerun" to lighting panels. ("Homerun" indicates the number of wires and conduit size from the last outlet box.)
- Use notes to simplify. For example, all wires shall be 2 #12 in 3/4" conduit unless otherwise indicated. Remember the information should be clear to insure proper illumination.

LIGHTING QUALITY

Illumination levels calculated by the lumen and point methods at best give only a "ballpark" estimate of the actual foot candle value to be realized in an installation. Many inaccuracies can be present, including differences between rated lamp lumen output and actual values, difficulty in predicting actual light loss factors, difficulty in predicting room surface reflectances, inaccurate CU information from a manufacturer, and non-rectangular shaped rooms.

Precise illumination levels are not critically important, however. Of equal importance to lighting quantity is lighting quality. Very few people can perceive a difference of plus or minus ten footcandles, but poor quality lighting is readily apparent to anyone and greatly affects our ability to comfortably "see" a task.

Of the many factors affecting the quality of a lighting installation, glare has the greatest impact on our ability to comfortably perceive a task. Figure 6-8 shows the two types of glare normally encountered.

DIRECT GLARE

Direct glare is often caused by a light source in the midst of a dark surface. Direct glare also can be caused by light sources, including

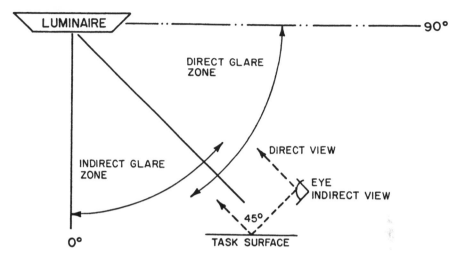

Figure 6-8. Direct and Indirect Glare Zones

sunlight, in the worker's line of sight. A rating system called Visual Comfort Probability (VCP) has been developed for assessing direct glare. The VCP takes into account fixture brightness at different angles of view, fixture size, room size, fixture mounting height, illumination level, and room surface reflectances.

Most manufacturers publish VCP tables for their fixtures. A VCP value of 70 or higher usually provides acceptable brightness for an office situation. Table 6-7 shows a typical VCP table.

INDIRECT GLARE

Indirect glare occurs when light is reflected off a surface in the work area. When the light bounces off a task surface, details of the task surface become less distinct because contrast between the foreground and background, such as the type on this page and the paper on which it is printed, is reduced. This is most easily visualized if a mirror is placed at the task surface and the image of a light fixture is seen at the normal viewing angle.

This form of indirect glare is called a veiling reflection because its effects are similar to those that would result if a thin veil were placed between the worker's eyes and the task surface. Veiling reflections can be reduced by:

Table 6-7

Room Size (in ft.)		Luminaires Lengthwise		Luminaires Crosswise	
		Ceiling Height (in feet)			
W	L	8.5	10.0	8.5	10.0
20 x	20	75	72	73	70
	30	75	72	73	70
	40	75	73	73	71
	60	75	73	72	71
30 x	20	78	75	77	73
	30	78	75	76	73
	40	77	75	75	72
	60	76	74	74	72
	80	76	74	73	72
40 x	20	81	78	80	77
	30	79	77	78	76
	40	78	77	76	75
	60	77	76	75	74
	80	77	75	74	73
	100	76	75	74	73
60 x	30	81	79	80	77
	40	79	78	78	76
	60	78	77	76	74
	80	77	76	75	74
	100	77	76	75	73
100 x	40	81	80	80	78
	60	80	78	78	76
	80	78	77	77	75
	100	78	76	76	74

**Wall Reflectance, 50%
Ceiling Cavity Reflectance, 80%
Floor Cavity Reflectance, 20%
Work Plane Illumination, 100fc

1) Orienting fixtures (or work surface) so that the light produced is not in the indirect glare zone. (Generally, to the side and slightly behind the work position gives the best results).

2) Selecting fixtures that direct the light above the worst veiling angles (generally 30° or greater). Note that in selecting fixtures to minimize indirect glare, care must be taken not to select fixtures that are a source of excessive direct glare.

NON-UNIFORM LIGHTING

The lumen method presented previously is useful for calculating average uniform illumination for an area, but illumination levels are for specific tasks in an area. By tailoring illumination levels to the various tasks in an area, significant energy savings are possible.

JOB SIMULATION-SUMMARY PROBLEM
SIM 6-5

(a) The Ajax Plant, SIM 2-6, Chapter 2, contains a workshop area with an area of 20' × 18'6".

For this area compute the number of lamps required, the space between fixtures, and the circuit layout. Use two 40-watt fluorescent lamps per fixture, 2900 lumens per lamp, light loss factor of .7, 110-volt lighting system, 20-watt ballast loss per fixture, and a fixture length of 2' × 4'. Use luminaire data from Table 6-8, ceiling height 20', and a desired footcandle level of 40.

Table 6-8. Coefficient of Utilization 20% Effective Floor Cavity Reflectance

Effective Ceiling Cavity Reflectance	80%			50%		
Wall Reflectance	50	30	10	50	30	10
RCR						
10	.33	.26	.22	.31	.26	.22
9	.43	.35	.27	.40	.35	.29
8	.58	.42	.35	.48	.42	.36
7	.58	.50	.42	.55	.48	.42
6	.64	.57	.49	.61	.54	.47
5	.72	.65	.59	.65	.60	.56
4	.77	.71	.64	.71	.65	.60
3	.82	.76	.70	.74	.69	.63
2	.87	.82	.77	.78	.74	.70
1	.91	.87	.83	.81	.78	.75

Spacing not to exceed 1 × Mounting Height

Analysis

The area of the workshop is 20' × 18'6".

Assume   hfc = 3

hcc = 3 Therefore,

hrc = 14

Assume   70% ceiling reflectance
          50% wall reflectance

The room cavity ratio is 7 and the effective ceiling cavity ratio pcc = 53.
Thus CU = .55.

$$\text{No. of Fixtures} = \frac{20' \times 18\text{-}1/2' \times 40}{2 \times 2900 \times .55 \times .70} = 7$$

Each 20 amp lighting circuit can provide power for up to 16 fixtures.

Layout Spacing
    3x = 20
     x = 6.6
    3y = 18–1/2'
     y = 6'2"

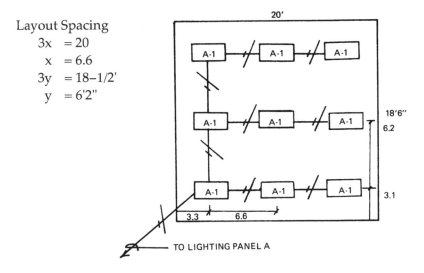

Note: With emphasis on energy conservation, a lighting layout using six
         fixtures may be preferable.

SUMMARY

Energy conservation is influencing lighting design. Increased
emphasis is being placed on minimizing lighting loads by using lamps and
luminaries that have high lumen outputs and coefficients of utilization.
Today's lighting systems incorporate switching and automatic control
devices which easily turn off lights when they are not required. Lighting
systems need to be analyzed on a first and operating cost basis to ensure

that the increasing energy costs are taken into account. Lighting system design must consider not only the quantity of illumination but also the quality of illumination. The choice of a luminaire and its location play an important part in comfortably perceiving a task. An awareness of the importance of quality lighting can result in a visual environment that is productive as well as energy efficient.

Chapter 7

Using Logic to Simplify Control Systems

Automated process plants are controlled by electrical hardware specified on elementary diagrams and designed by electrical engineers. Standard contactors, timing devices, relays, and other provide control functions. In this chapter an approach for analyzing and designing elementary diagrams is developed. Learn the logic language, a way for communicating.

ELECTRICAL HARDWARE
USED FOR CONTROL

Electrical hardware used for control includes:
- *Relays*—Devices that close when energized. Contacts physically located on the relay will either open or close when the relay coil is energized.
- *Timers and time delay relays*—Devices whose contacts close or open after a preset time.
- *Push buttons*—Devices which are used to actuate a control system, i.e. stop-start push buttons.
- *Programmers*—Devices with contacts that open and close in a preset sequence.

Plug-in type relays and prepackaged solid-state components are widely used.

SYMBOLS

Symbols commonly used in electrical control schematics are illustrated in Figure 7-1. These symbols are based on the Joint Industrial Council (JIC) Standards.

THE ELECTRICAL ELEMENTARY (SCHEMATIC)

Figure 7-2 illustrates a typical electrical elementary diagram. Notice that line identifications on the left are used as references to locate relay contacts on the right. For example, relay R_1 has a normally open contact on line six.

Steps for Analyzing Electrical Control Circuits
- Look at one line of operation at a time.

- Trace a path of power from left to right. Every contact to the left of the electrically operated line must be closed for the device to operate.

- Each line is identified with a consecutive number on the left.

- Numbers at the right of the line, next to a relay, show on what lines the device has contacts. Underlined numbers indicate a normally closed contact.

Simple Control Schemes
In order to control a motor, a starter is required. A typical motor elementary diagram, including the power portion, is illustrated in Figure 7-3. HCA represents the holding coil or contactor of the starter. HCA is simply a relay which has contacts capable of interrupting power to the associated motor.

If the stop-start push button were located locally, three wires, numbers 1, 2, 3 would need to run from the field to the MCC.

The control voltage in Figure 7-4 is 480 volts, but if the circuit had a solenoid valve, limit switch, etc., 110 volts would be required. (Control devices are usually rated for 110 volts.)

Lockouts
Two typical lockout schemes are illustrated in Figure 7-4. Scheme "A" illustrates the case where Motor "B" can not be started unless Motor "A" is

Figure 7-1. Electrical Symbols

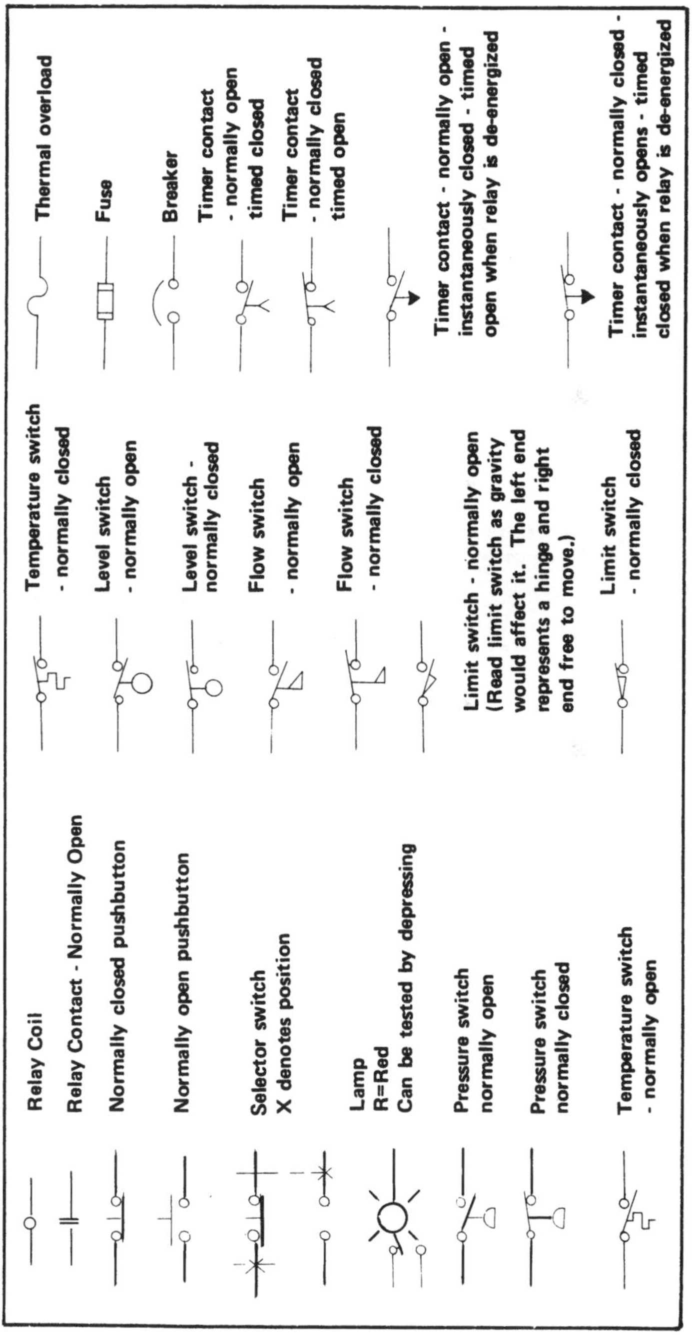

Relay Coil

Relay Contact - Normally Open

Normally closed pushbutton

Normally open pushbutton

Selector switch
X denotes position

Lamp
R=Red
Can be tested by depressing

Pressure switch
normally open

Pressure switch
normally closed

Temperature switch
- normally open

Temperature switch
- normally closed

Level switch
- normally open

Level switch -
normally closed

Flow switch
- normally open

Flow switch
- normally closed

Limit switch - normally open
(Read limit switch as gravity
would affect it. The left end
represents a hinge and right
end free to move.)

Limit switch
- normally closed

Thermal overload

Fuse

Breaker

Timer contact
- normally open
timed closed

Timer contact
- normally closed
timed open

Timer contact - normally open -
instantaneously closed - timed
open when relay is de-energized

Timer contact - normally closed -
instantaneously opens - timed
closed when relay is de-energized

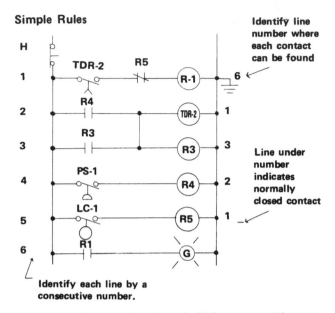

Figure 7-2. Format of an Electrical Elementary Diagram

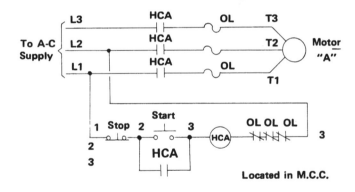

Stop Start - Full Voltage
Non-reversing Motor

Figure 7-3. Elementary Diagram for a FVNR Motor

running. If Motor "A" stops, Motor "B" will be locked out and stop.

Scheme "B" illustrates the case where once Motor "B" is running it does not matter if Motor "A" stops. The permissive lockout is only required to start Motor "B."

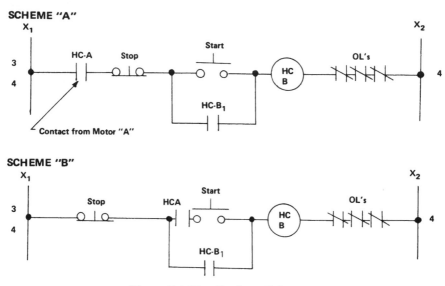

Figure 7-4. Two Lockout Schemes

Reversing Motors

To change the direction of rotation of a motor, it is necessary to switch any two of the motor leads. Figure 7-5 illustrates the elementary for a reversing motor. Interlocks prevent activating both directions (at once).

Two-speed Motors

To change the speed of the motor requires changing the effective number of poles. This is accomplished by either using a motor with two separate windings (TSSW) or a motor whose windings are taped (TSCP) so that they can be connected in two ways. Six power leads are required to change the windings. Figure 7-6 illustrates an elementary diagram for a two-speed motor with separate windings. The control portion of the elementary is similar to that of a reversing motor (Figure 7-5), except that six overloads are required. Interlocks prevent activating both speeds (at once).

DESCRIPTION OF OPERATION AND LOGIC DIAGRAMS

To describe how a process operates it is necessary to establish a logic diagram or description of operation. From this description the electrical schematic is designed.

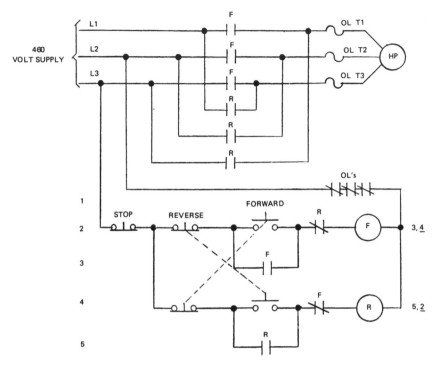

Figure 7-5. Elementary Diagram for a FVR Motor

SIM 7-1

From the description of operations, draw an electrical schematic.
The HVAC design is as follows:

(a)   When exhaust fan #1 is operated, damper EP valve is energized (electric to pneumatic).

(b)   A PE (pneumatic to electric) switch prevents #2 fan from starting.

Answer

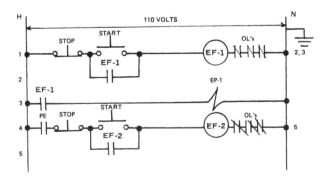

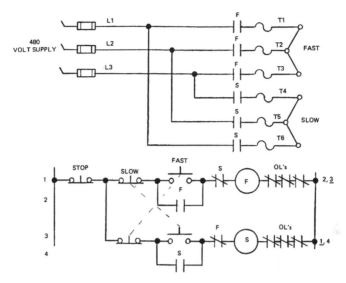

Figure 7-6. Elementary Diagram for a Two-speed Motor

SIM 7-2

Draw elementaries for each scheme.

(a) When level switch LCH #1 in Tank "A" reaches "high" level, Pump #3 is started. Pump will run for ten minutes before it will turn off automatically.

(b) Pump #4, which feeds Tank "A," can be started manually and will automatically stop when high level occurs (LCHA).

(c) There are three valves that are used at a manifold station. When Valve "A" is energized, Valve "C" cannot be energized. When Valve "B" is energized, Valves "A" and "C" cannot be energized. Once Valve "C" is energized, Valves "A" and "B" have no effect.

Answer

It should be noted that several electrical schematics may all be correct but look different. Another point is that several people may interpret the above descriptions differently. One possible solution is illustrated on page 208.

LOGIC DIAGRAMS AND LOGIC CONTROLLERS

Logic diagrams use symbols to describe the operation, whereas the elementary diagram uses more words. The logic diagram is an

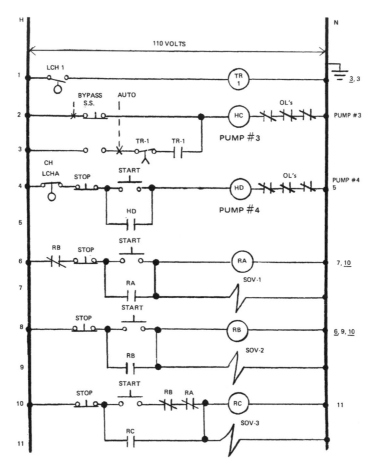

effective tool to convey process requirements, since it is understood by all engineers. To describe a process, the logic diagram uses symbols for the three words "And," "Or," and "Not." Since any logic can be conveyed with these three words, it offers a simple, exact means to describe process requirements. Figure 7-7 illustrates symbols commonly used in logic. These symbols are combined to describe a process. It is possible to design a complete electrical control scheme by using a programmable logic controller (PLC) diagram. A PLC is basically a computer that can be easily programmed for any process. Unlike relays, they are not hard-wired, so changes in operation can be readily made by just reprogramming the unit. It is possible to use the input directly from the logic diagram to program the PLC without changing the elementary diagram as an intermediate step.

Logic Functions	NEMA Symbol	Description
AND		A device which produces an output only when every input is present—represents contacts in series.
OR		A device which produces an output when one input (or more) is present—represents contacts in parallel.
NOT		A device which produces an output only when the input is absent — represents a normally closed contact.
ON DELAY TIMER		A device which produces an output following a definite time delay after its input is applied.
OFF DELAY TIMER		A device whose output is removed following a definite time delay after its input is removed.

Figure 7-7. Logic Symbols

Thus, logic diagrams can save design time when used with programmable devices. They are also useful in the development of conventional control schemes, since they offer a visual tool that can be understood not only by the electrical engineer, but by the process and mechanical engineer as well. This means that prior to the start of the electrical elementaries and interconnecting drawings the process can be resolved. This in itself can minimize costly design changes.

HOW TO ANALYZE LOGIC DIAGRAMS

To analyze a logic diagram it is necessary to determine the various inputs required to actuate the logic gate. When time delays are incorporated into the design, it is necessary to determine the "state" of the process at various time periods. One method to accomplish this is to place a "1" or a "0" after each logic symbol.

A "1" indicates a signal present, and a "0" indicates the absence of a signal. Always try to establish the initial state of the circuit. The logic elements can be combined in any order to describe the electrical circuit.

SIM 7-3

SWITCH A

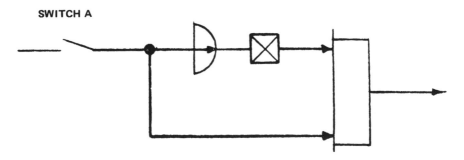

What will happen when Switch "A" is closed?
Draw an elementary.

Answer

Step I Identify initial operating and final state of logic.
    Comments—The above logic shows how various elements can be combined. Note that two signals from switch "A" are used. One goes

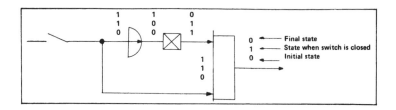

through a "Time Delay" and "Not" gate and the other is an instantaneous signal fed directly into an "And" gate. Both signals are required to give a pulse after the switch is closed.

Step 2   Look at output—a pulse when equipment is activated.

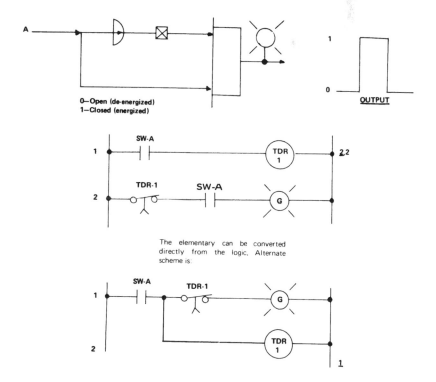

SIM 7-4

What will happen when SE.1 is closed?
Draw an elementary diagram.

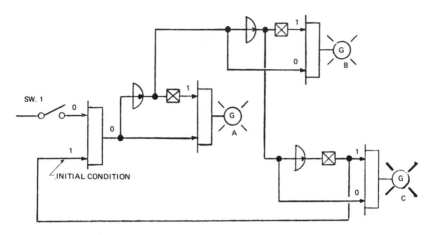

Answer

How to Analyze Logic

(a) Determine the initial condition with switch open.

(b) By placing "1s" and "0s" on the above diagram, determine the operation when switch is closed.

(c) Always check the final state to see if logic resets.

(See illustration on opposite page)

The above elementary and associated logic represent sequential pulses when the switch is closed.

SIM 7-5

From the description of operation, draw a logic diagram and an elementary.

1. In order for the compressor (C-1) to operate, the oil oump (P-2) should be running for five minutes. An interlock should be provided so that if an operator pressed the start button, he must keep the button depressed until the above is accomplished.

2. After the compressor is running, the auxiliary oil pump will be manually stopped by the operator.

3. If low pressure should occur during compressor operation, auxiliary oil pump will automatically start. (Stop same as 2.)

4. When the compressor is shut down, auxiliary oil pump will automatically come on and automatically stop after five minutes.

5. If high pressure occurs, the compressor will automatically stop.

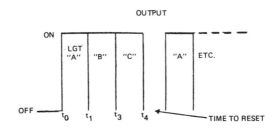

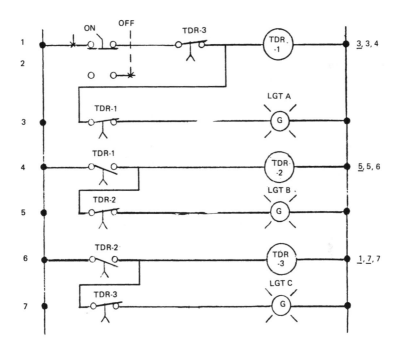

6. A manual STOP-START push button for the auxiliary oil pump is provided.

7. A manual STOP-START push button for the compressor is provided.

Answer

Analysis

Step 4 is identical with the scheme of a pulse when a signal goes off. Since the compressor is the simpler scheme, first draw the compressor logic.

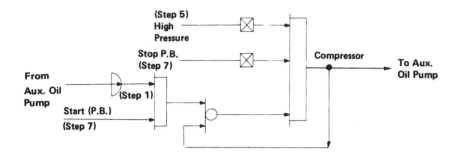

Compressor Logic

In the auxiliary oil pump scheme, the motor can be manually started or started through low pressure. Since Step 4 also automatically stops the motor, this step must be drawn independently of the seal in contact.

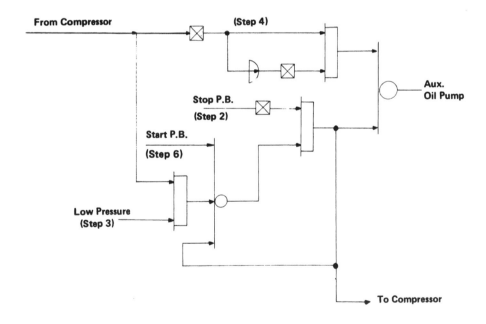

The two basic schemes can be combined, as illustrated on the following page.

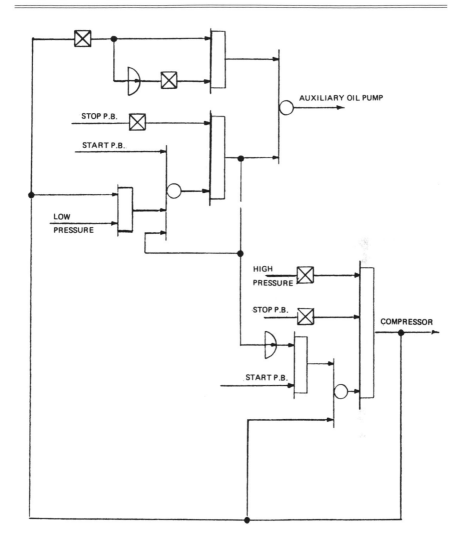

INTERCONNECTION DIAGRAMS

To show the electrician how to wire from an elementary diagram, a physical arrangement referred to as an interconnection diagram is made. Vendor's diagrams show how the terminal blocks are arranged and are used. Each wire on the elementary is assigned a wire number. All devices are connected internally to the terminal blocks for each panel. The electrician needs only to connect the terminals together with the control cable.

Development of Elementary

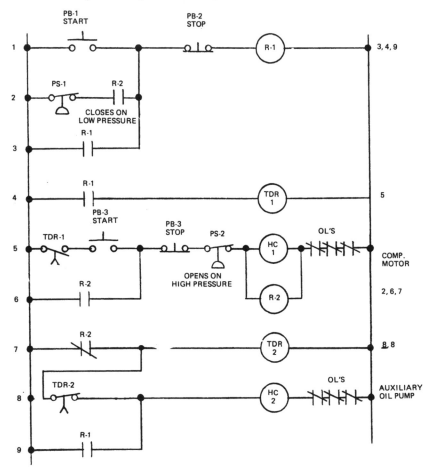

SIM 7-6

Make an interconnection diagram for the elementary on the next page. Wire numbers and terminal numbers have the same designation as indicated.

Push buttons are located on Push button Panel No.1. Relays are located on Relay Panel No. 1. Solenoid valves are located locally; 110 volts source is from MCC No. 1.

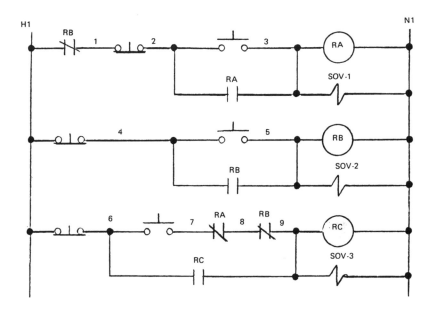

Answer

(See illustration on page 218)

JOB SIMULATION—SUMMARY PROBLEM

JOB 6

(a) Draw an electrical control scheme for CF-3, C-16 and TP-5. When CF-3 is reversing, the transfer pump TP-5 can not be operated. The conveyor C-16 will not be able to operate in the forward cycle of the centrifuge CF-3. Push buttons for C-16 and CF-3 are located on Panel No. 1. Push buttons for TP-5 are locally mounted.

(b) Draw a typical stop-start scheme for motors CTP-6, HF-10, HF-11, BC-13 and SC-19. These motors can be started locally and at the MCC.

(c) From the elementaries developed in (a) and (b) of this problem, and the elementary illustrated, complete the control portions of the conduit and cable schedule.

SIM 7-6 Answer:

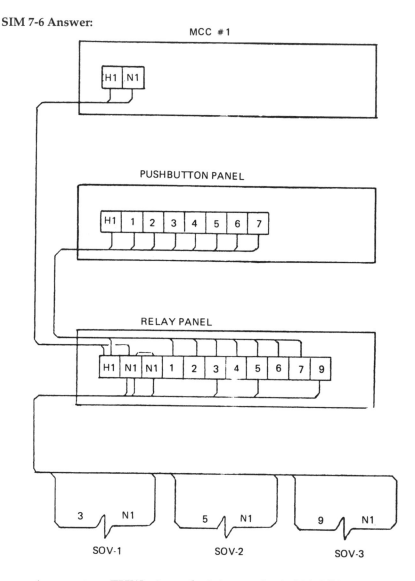

MCC #1

PUSHBUTTON PANEL

RELAY PANEL

SOV-1   SOV-2   SOV-3

Assume type THW wire and minimum size is #14. All interconnections are made at the MCC, and field devices are located near each other. Common wires will be jumped locally. Designate conduits as follows:

C-P1   From MCC No.1 to Panel No. 1.
C-R1   From MCC No.1 to Relay Panel No. 1.
C-L1   From MCC No.1 to local devices

(NOTE: local push buttons are excluded since those cables were sized in Job 5).

The motor list is based on Job 1, Chapter 2.
Illustrated elementary [see (c)] for motors AG-1, FP-4, CT-9, UH-12 and RD-22.

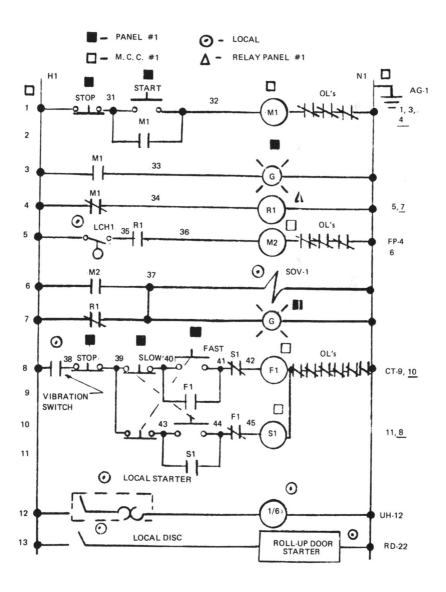

Analysis

(a) and (b)

■ PANEL #1   ◉ LOCAL

□ M.C.C. #1   △ RELAY PANEL #1

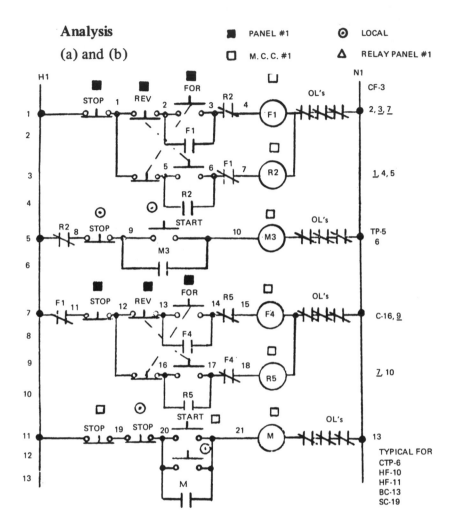

(c)

Conduit Designation	No. Conductors	Wire Size	Conduit Size	Wire Numbers in Each Conduit
C-P1	20	#14	1¼″	H1, N1, 2, 3, 5, 6, 11, 13, 14, 16, 17, 31, 32, 33, 37, 38, 40, 41, 43, 44
C-R1	6	#14	¾″	H1, N1, 34, 35, 36, 37
C-L1	5	#14	¾″	H1, N1, 35, 37, 38

SUMMARY

Many times it is difficult to obtain a description of operation from the process, project or mechanical engineer. In many plants the elementary diagram is the only document that describes how a process works. Usually only the electrical engineer understands the elementary diagram so it is difficult to ensure that the elementary diagram, meets the functional requirements. The verbal description that is the input to the elementary diagram can be easily misinterpreted. Since all disciplines understand the simple logic elements, logic diagrams can be initiated and agreed upon. The electrical engineer can then use the logic diagram as the basis for his design. By representing the circuit in terms of a logic diagram, the electrical engineer can use Boolean algebra to simplify the elementary control.

Solid-state logic or programmable controllers can be obtained from the logic diagram. The logic elements form a common language permitting the process, mechanical, project and electrical engineers to create a complex elementary control scheme. The individual who understands the fundamentals of the logic elements can easily apply this important tool in developing complex industrial and power elementary control schemes.

Programmable Logic Controller Diagram Example

**Motor with Forward and Reverse Control
with Time-Delay from One Direction to the Other
Logic Controller Ladder Form with Logic Mnemonics**

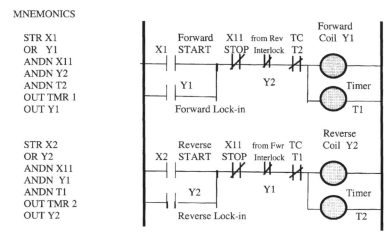

MNEMONICS

STR X1	Forward X11 from Rev TC → Forward Coil Y1
OR Y1	X1 START STOP Interlock T2
ANDN X11	
ANDN Y2	
ANDN T2	Y1 Y2 → Timer T1
OUT TMR 1	Forward Lock-in
OUT Y1	
STR X2	Reverse X11 from Fwr TC → Reverse Coil Y2
OR Y2	X2 START STOP Interlock T1
ANDN X11	
ANDN Y1	
ANDN T1	Y2 Y1 → Timer T2
OUT TMR 2	Reverse Lock-in
OUT Y2	

Note must press STOP to change from one direction to the other and TC ≡ timed closed

Chapter 8

Applying Process Controllers and Electronic Instrumentation

The basic requirement for developing electrical schematics is to understand the process. The logic diagram is of particular use in describing a process when the activating input is related to an on-off signal. In the past, the hardware used to accomplish this logic was a combination of heavy-duty machine tool relays. In use today are solid-state programmable controllers, particularly where the process steps repeat continuously. In addition, the current trend is to control process variables (i.e., temperature, pressure, flow level) by electronic instrumentation loops. Thus it is important to understand how programmable controllers and instrumentation are applied.

PROGRAMMABLE LOGIC CONTROLLER

The programmable logic controller (PLC) developed out of the needs of the automotive industry. The industry required a control unit which was easily changed in the plant, easily maintained/repaired, highly reliable, small, capable of outputting data to a central data collection system, and competitive in cost to relay and solid-state panels. Out of these requirements developed the first series of programmable controllers. These units were initially designed to accept 115-volt AC inputs, and to output 115 volts AC, with 2 ampere signals capable of actuating solenoid valves, motors, etc. The memory was capable of expansion to a maximum of 4000 words.

Today the PLC is far more flexible and reliable than the earlier generations, and its use has had far-reaching implications to most industrial plants. Let's examine some of the features of most models.

The basic unit described in Figure 8-2 consists of:

Various Programmable Logic Controllers

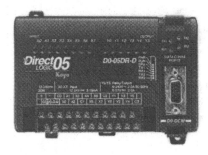

Figure 8-1a. "Brick" Logic Controller

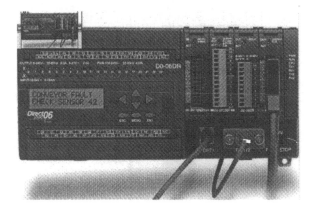

Figure 8-1b. Logic Controller with a Few Modules

Figure 8-1c. Logic Controller with Rack and Numerous Modules

Figures above used with permission of Automation Direct

Mnemonics

STR X1
OR X2
ANDN Y2
OUT Y1
STR X1
AND X2
OUT Y2

Corresponding Ladder Logic Diagram

Combined Local and General Alarm System

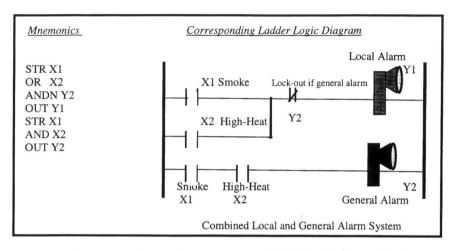

Figure 8-1d. Automation Direct Programmable Logic Controller Example Basics

Boolean Logic	*Corresponding Mnemonics*
X1 = Y1	STR X1, OUT Y1
NOT X2 = Y2	STRN X2, OUT Y2
X3 AND X4 = Y3	STR X3, AND X4, OUT Y3
X5 OR X6 = Y4	STR X5, OR X6, OUT Y4
X11 AND NOT X12 = Y5	STR X11, ANDN X12, OUT Y5

Figure 8-1e. Automation Direct Programmable Logic Controller Mnemonics and Ladder

- Input module
- Memory
- Processor
- Output module
- Programming auxiliary

Input Module

The input module usually accepts AC or DC signals from remote devices such as pushbuttons and switches. These signals are then converted to DC levels, filtered and attenuated for use by the processor.

Processor

This module is the working portion of the PLC. All input and outputs are continuously monitored. The status of the input is compared against an established program, and instructions are executed to the various outputs. The control function of the processor identifies the memory core locations to be addressed. An internal timing device determines the required sequence to fetch and execute instructions.

Memory

The information stored in the memory relates to how the input-output data should be processed. Information is usually stored on magnetic cores.

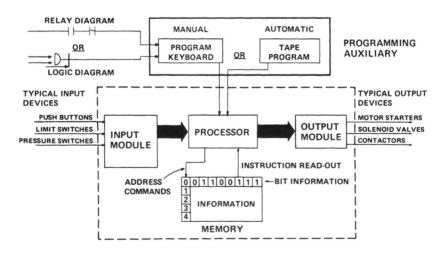

Figure 8-2. Components of a Programmable Controller

Output Module

The output module provides the means to command external machine devices. Output loads are usually energized through triac AC switches or reed contacts.

Many units have input and output lights on the unit that indicate the status of the remote devices.

Programming the PLC

Two common typical methods for programming are:

Use of External Hand-held Programmer Keyboard

The handl-held programmer keyboard may use logic symbols or standard elementary symbols. In either case, the logic or elementary is directly programmed into the unit. Refer to Figure 8-3.

Use of a PC with Software

A program can be stored in a personal computer (PC), allowing the program to be developed in one location and transferred to the site. Additionally, this allows a backup copy of the program to be available if the system goes down.

Another feature of programming is in elementary simplification and simulation. If a PC is used for programming, the information can be used

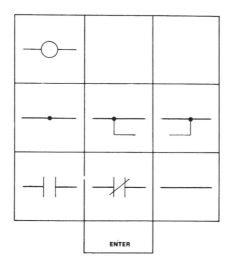

Figure 8-3. Typical Keyboard for Programmable Controller

in a standard elementary simplification program. Any redundancies in the logic can be quickly found. The PC can also be used in conjunction with a simulation program. For any input, the corresponding output can be determined without the actual output being energized. To change sequence of operation, the memory is simply reprogrammed.

Capabilities

The PLC is capable of performing the function of hard wired solid-state systems or relay systems. The basic functions are:

- **Logic Gates**—For a review of logic, see Chapter 7.

- **Timers**—Either on-delay or off-delay timers. Typical ranges .1 to 99.9 seconds.

- **Counters**—Used to count input status changes at a rate of 20 per second. Counter size is 999 counts.

- **Latches**—Simulate electro-mechanical relay.

- **Shift Registers**—Provide the ability to simultaneously remember the state of several pieces as they move through the manufacturing process.

- **Maintenance**—Many units feature plug-in modules and a preprogrammed diagnostic tape so problems can be quickly identified and corrected.

Applications

There are many PLCs available. Each unit may have slightly different features but all offer the versatility of a non-hard wired controller. The size of the input-output and logic capabilities also varies.

The PLC should always be considered since it offers:

- Complete flexibility to modify or expand controls when production needs change.
- A highly reliable solid state control that minimizes downtime.
- The possible reduction of engineering design time (no need to check back of panel wiring, etc.).
- A system compatible with computer simplification and simulation techniques.
- Fast diagnostics and maintenance.

INSTRUMENTATION

Modern process control mostly uses electronic instrumentation. In the past, pneumatic instrumentation dominated the market, but today most control loops are electronic.

Closed Loop

To control a process requires a closed loop system. In the example of controlling level, if the level is too high the level is reduced. Figure 8-4 shows a typical closed loop system.

Where:

C   is the controlled variable
R   is the reference or set point
E   is the error or deviation
M   is the variable manipulated by the controller.

Figure 8-5 shows a typical process control diagram illustrating level control.

In this process, the level transmitter (LT-1) sends the electronic signal to the level controller recorder (LCR-1). If the level is too low, control valve CV-1 is throttled. On the other hand, if the level is too high, CV-1 is opened. Thus, the process is controlled. This example illustrates the fundamental elements of a control loop, namely:

Input (transmitter)
Controller
Output (control valve).

Signal

In electronic instrumentation the standard signal used to convey

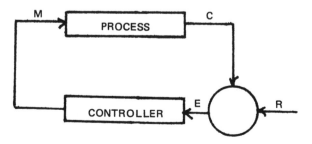

Figure 8-4. Typical Closed Loop System

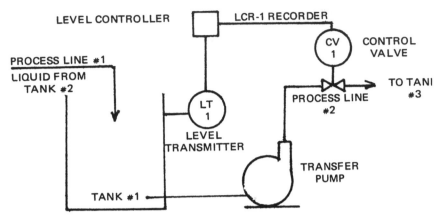

Figure 8-5. Typical Level Control Process

process variables is 4-20 ma DC.

Figure 8-6 illustrates the process variable in percentage as a function of the corresponding electronic signal.

This means that a full-scale level reading corresponds to 100%. In this example an 8 ma DC signal from LT-1 would correspond to a 25% reading on the level controller.

Control Loop

The basic control circuitry for electronic instrumentation is illustrated in Figure 8-7. This figure illustrates a typical series circuit for the process of Figure 8-5. Each receiving element contains an input impedance.

In a series network, the impedance of each element must be carefully matched.

DIRECT DIGITAL CONTROL

Process controllers are beginning to be used to make not only logic decisions but also control loop decisions. Figure 8-8 shows a process controller configuration to accomplish the process of Figure 8-5. Such a configuration is often called direct digital control (DDC), since the control action is determined by the program within the controller. DDC offers an easy way of changing control strategies without rewiring or recalibrating.

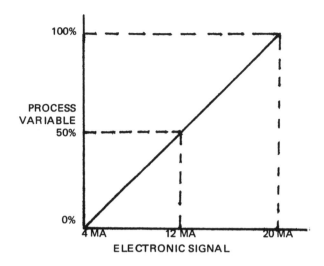

Figure 8-6. Process Variable vs. Electronic Signal

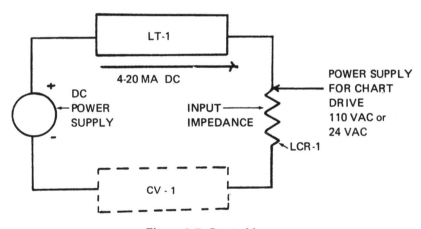

Figure 8-7. Control loop

Additionally, DDC requires much less calibration than do pneumatic and electronic controllers.

Types of Control Instrumentation

The several common types of control variables are:

- Level
- Flow

- Pressure
- Temperature

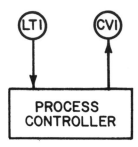

Figure 8-8. Direct Digital Control

Level

As indicated in Figure 8-5, level is controlled by CV-1. Level could also be controlled by a control valve on process line No.1. Two common types of level transmitters are the float type and differential pressure type.

Flow

As indicated in Figure 8-9, flow is essentially controlled the way level is controlled. The signal from FT-1 is sent to FC-1, and control valve CV-1 is either opened or closed. Two common types of flow transmitters are the magnetic flow meter and the orifice meter type.

Magmeter Type

The magnetic flowmeter is the more expensive of the two. The magnetic flow meter uses the same principle of operation as a tachometer or generator. As indicated in Figure 8-10, the fluid flowing acts as the conductor, while the pipe is located in the magnetic field caused by the field coils. The electrodes are mounted in a plane at right angles to the magnetic

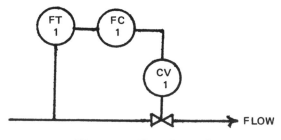

Figure 8-9. Flow Control

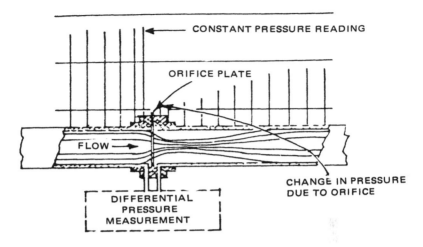

Figure 8-10. Magnetic Flow Meter

field and act like brushes of a generator. Thus the voltage induced by the moving fluid is brought out by leads for external measurement. The induced voltage can be converted by the transmitter to direct current. For magmeters to work properly, the fluid must conduct electricity, and the tube must be liquid full.

Orifice Meter Type

As liquid flows through a restriction in a pipe orifice plate, a change in pressure occurs as illustrated in Figure 8-11. The flow rate is calculated by measuring the differential pressure across the orifice plate.

Pressure

Pressure control is used to maintain a specified pressure or is used as an indirect measurement of level or flow, as indicated previously.

Temperature

The thermocouple is one of the most frequently used methods for measuring temperatures between 500 and 1500°C. Strictly speaking, the thermocouple and its associated control do not fall into the typical electronic instrumentation category.

The operation of a thermocouple is based on the principle that an electromotive force (emf) is developed when two different metals come in contact. The emf developed is dependent on the metals involved and

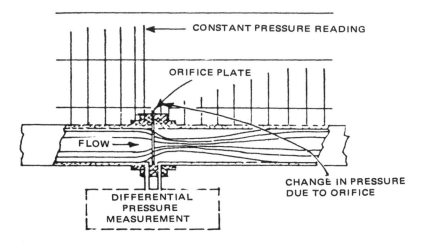

Figure 8-11. Orifice Flow Meter

the *temperature* of the junctions. The emf signal developed is expressed in millivolts. Due to the nature of the thermocouple, wire splices should be avoided. Two types of measuring circuits are used in conjunction with thermocouples.

Figure 8-12 illustrates a typical thermocouple circuit using a Galvanometer.

Figure 8-13 illustrates a typical thermocouple circuit using a potentiometer.

The resistance temperature device (RTD) is a common device used in measuring temperatures. The RTD resistance varies with temperature; consequently when it is installed in a circuit the resulting voltage drop is proportional to the temperature.

WIRING METHODS

The type of cable and the installation of instrument signals are based on the reduction of noise. In general, instrument cables are routed away from noise sources such as power cables, motors, generators, etc. Twisted control cables are used to reduce magnetic noise pickup from a nearby noise source.

Table 8-1 summarizes instrumentation wiring methods.

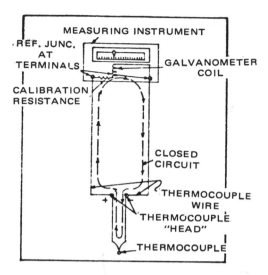

Figure 8-12. Basic Galvanometer Thermocouple Circuit

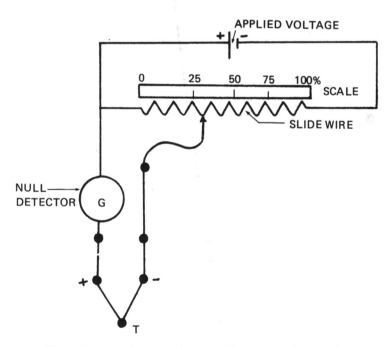

Figure 8-13. Basic Potentiometer Thermocouple Circuit

Table 8-1. Instrumentation Wiring Methods.

	RECEIVING ELEMENT DESCRIPTION	
TRANSMITTER TYPE	Torque type	Bridge or potentiometer type
	current converted directly to a torque to move a chart recorder or pointer, i.e., pyrometer, Galvanometer	input signal is compared with a standard voltage and amplified to drive a chart or recorder to a balanced or null position
THERMOCOUPLE	twisted pair nonshielded	twisted pair shielded
	NOTE: lead wire must be the same material as the thermocouple, i.e., iron constantan	
MAGNETIC FLOW METERS		twisted pair shielded
PNEUMATIC CURRENT TRANSDUCERS, DIFFERENTIAL PRESSURE FLOW METERS	twisted pair nonshielded	*twisted pair nonshielded *NOTE: for a high noise environment use twisted pair shielded.

Chapter 9

Protective Relaying for Power Distribution Systems

Due to possible equipment failure or human error, it is necessary to provide protection devices. These devices minimize system damage and limit the extent and duration of service interruption when failure occurs. The main goal of protection coordination is to isolate the affected portion of the system quickly while at the same time maintaining normal service for the remainder of the system. In other words, the electrical system must provide protection and selectivity to ensure that the fault is minimized while other parts of the system not directly involved are held in until other protective devices clear the trouble. Protective devices such as fuses and circuit breakers have time current characteristics that determine the time it takes to clear the fault. In the case of circuit breakers it is possible to adjust the characteristics, while fuse characteristics are nonadjustable.

Protective relays are another way of achieving selective coordination and are required to operate power breakers above 600 volts. By definition, a protective relay is a device which when energized by suitable currents, voltages, or both, responds to the magnitude and relationships of these signals to indicate or isolate an abnormally operating condition. These relays have adjustable settings and can be used to actuate the opening of circuit breakers under various fault conditions.

In Chapter 4, the concept of the one-line diagram was introduced. In order to complete the power distribution system, it is necessary to show on the one-line diagram the protective relaying required and the breakers affected. To properly set the protective devices it is necessary to know the fault currents that occur at various portions of the system. This chapter will illustrate typical applications of protective devices.

THE OVERCURRENT RELAY (Device 51)

Every protective relay has an associated number. Some of the standard designations are listed in Table 9-1. The overcurrent relay is designated as Device 51. This relay is used for overcurrent protection and is current sensitive. The one-line diagram utilizing the overcurrent relay is shown in Figure 9-1.

Overcurrent relays are available with inverse, very inverse, and extremely inverse time current characteristics. The very inverse time current characteristic is the frequent choice when detailed system information is not available. The very inverse characteristic is most likely to provide optimum circuit protection and selectivity with other system protection devices. When coordinating with fuses it may be necessary to use the extremely inverse characteristic.

Figure 9-2 illustrates the characteristics of electromechanical overcurrent protective relays. The overcurrent relays are current sensitive and require a seal-in contact to keep them energized after activation.

Table 9-1. Protective Device Numbering and Functions

DEVICE NUMBER	DEFINITION AND FUNCTION
1	master element is the initiating device, such as a control switch, voltage relay, float switch, etc., which serves either directly, or through such permissive devices as protective and time-delay relays to place an equipment in or out of operation.
2	time-delay starting, or closing, relay is a device which functions to give a desired amount of time delay before or after any point or operation in a switching sequence or protective relay system, except as specifically provided by device functions 62 and 79 described later.
3	checking or interlocking relay is a device which operates in response to the position of a number of other devices, or to a number of predetermined conditions in an equipment to allow an operating sequence to proceed, to stop, or to provide a check of the position of these devices or of these conditions for any purpose.
4	master contactor is a device, generally controlled by device No. 1 or equivalent, and the necessary permis-

Table 9-1. Protective Device Numbering and Functions (*Continued*)

DEVICE NUMBER	DEFINITION AND FUNCTION
4 (con't.)	sive and protective devices, which serves to make and break the necessary control circuits to place an equipment into operation under the desired conditions and to take it out of operation under other or abnormal conditions.
5	**stopping device** functions to place and hold an equipment out of operation.
6	**starting circuit breaker** is a device whose principal function is to connect a machine to its source of starting voltage.
7	**anode circuit breaker** is one used in the anode circuits of a power rectifier for the primary purpose of interrupting the rectifier circuit if an arc back should occur.
8	**control power disconnecting device** is a disconnecting device—such as a knife switch, circuit breaker or pullout fuse block—used for the purpose of connecting and disconnecting, respectively, the source of control power to and from the control bus or equipment. **note:** Control power is considered to include auxiliary power which supplies such apparatus as small motors and heaters.
9	**reversing device** is used for the purpose of reversing a machine field or for performing any other reversing functions.
10	**unit sequence switch** is used to change the sequence in which units may be placed in and out of service in multiple-unit equipments.
11	Reserved for future application.
12	**over-speed device** is usually a direct-connected speed switch which functions on machine overspeed.
13	**synchronous-speed device,** such as a centrifugal speed switch, a slip-frequency relay, a voltage relay, an undercurrent relay or any type of device, operates at approximately synchronous speed of a machine.
14	**under-speed device** functions when the speed of a machine falls below a predetermined value.

/more/

Table 9-1. Protective Device Numbering and Functions (*Continued*)

DEVICE NUMBER	DEFINITION AND FUNCTION
15	**speed or frequency, matching device** functions to match and hold the speed or the frequency of a machine or of a system equal to, or approximately equal to, that of another machine, source or system.
16	Reserved for future application.
17	**shunting, or discharge, switch** serves to open or to close a shunting circuit around any piece of apparatus (except a resistor), such as a machine field, a machine armature, a capacitor or a reactor. note: This excludes devices which perform such shunting operations as may be necessary in the process of starting a machine by devices 6 or 42, or their equivalent, and also excludes device 73 function which serves for the switching of resistors.
18	**accelerating or decelerating device** is used to close or cause the closing of circuits which are used to increase or to decrease the speed of a machine.
19	**starting-to-running transition contactor** is a device which operates to initiate or cause the automatic transfer of a machine from the starting to the running power connection.
20	**electrically operated valve** is a solenoid- or motor-operated valve which is used in a vacuum, air, gas, oil, water, or similar, lines. note: The function of the valve may be indicated by the insertion of descriptive words such as "Brake" or "Pressure Reducing" in the function name, such as "Electrically Operated **Brake** Valve."
21	**distance relay** is a device which functions when the circuit admittance, impedance or reactance increases or decreases beyond predetermined limits.
22	**equalizer circuit breaker** is a breaker which serves to control or to make and break the equalizer or the current-balancing connections for a machine field, or for regulating equipment, in a multiple-unit installation.
23	**temperature control device** functions to raise or to lower the temperature of a machine or other apparatus,

/more/

Table 9-1. Protective Device Numbering and Functions (*Continued*)

DEVICE NUMBER	DEFINITION AND FUNCTION
23 (con't.)	or of any medium, when its temperature falls below, or rises above, a predetermined value.
	note: An example is a thermostat which switches on a space heater in a switchgear assembly when the temperature falls to a desired value as distinguished from a device which is used to provide automatic temperature regulation between close limits and would be designated as 90T.
24	Reserved for future application.
25	**synchronizing, or synchronism-check, device** operates when two a-c circuits are within the desired limits of frequency, phase angle or voltage, to permit or to cause the paralleling of these two circuits.
26	**apparatus thermal device** functions when the temperature of the shunt field or the armortisseur winding of a machine, or that of a load limiting or load shifting resistor or of a liquid or other medium exceeds a predetermined value; or if the temperature of the protected apparatus, such as a power rectifier, or of any medium decreases below a predetermined value.
27	**undervoltage relay** is a device which functions on a given value of undervoltage.
28	Reserved for future application.
29	**isolating contactor** is used expressly for disconnecting one circuit from another for the purposes of emergency operation, maintenance, or test.
30	**annunciator relay** is a nonautomatically reset device which gives a number of separate visual indications upon the functioning of protective devices, and which may also be arranged to perform a lockout function.
31	**separate excitation device** connects a circuit such as the shunt field of a synchronous converter to a source of separate excitation during the starting sequence; or one which energizes the excitation and ignition circuits of a power rectifier.
32	**directional power relay** is one which functions on a desired value of power flow in a given direction, or upon reverse power resulting from arc back in the anode or cathode circuits of a power rectifier. */more/*

Table 9-1. Protective Device Numbering and Functions (*Continued*)

DEVICE NUMBER	DEFINITION AND FUNCTION
33	**position switch** makes or breaks contact when the main device or piece of apparatus, which has no device function number, reaches a given position.
34	**motor-operated sequence switch** is a multi-contact switch which fixes the operating sequence of the major devices during starting and stopping, or during other sequential switching operations.
35	**brush-operating, or slip-ring short-circuiting, device** is used for raising, lowering, or shifting the brushes of a machine, or for short-circuiting its slip rings, or for engaging or disengaging the contacts of a mechanical rectifier.
36	**polarity device** operates or permits the operation of another device on a predetermined polarity only.
37	**undercurrent or underpower relay** is a device which functions when the current or power flow decreases below a predetermined value.
38	**bearing protective device** is one which functions on excessive bearing temperature, or on other abnormal mechanical conditions, such as undue wear, which may eventually result in excessive bearing temperature.
39	Reserved for future application.
40	**field relay** is a device that functions on a given or abnormally low value or failure of machine field current, or on an excessive value of the reactive component of armature current in an a-c machine indicating abnormally low field excitation.
41	**field circuit breaker** is a device which functions to apply, or to remove, the field excitation of a machine.
42	**running circuit breaker** is a device whose principal function is to connect a machine to its source of running voltage after having been brought up to the desired speed on the starting connection.
43	**manual transfer or selector device** transfers the control circuits so as to modify the plan of operation of the switching equipment or of some of the devices.

/more/

Table 9-1. Protective Device Numbering and Functions (*Continued*)

DEVICE NUMBER	DEFINITION AND FUNCTION
44	**unit sequence starting relay** is a device which functions to start the next available unit in a multiple-unit equipment on the failure or on the non-availability of the normally preceding unit.
45	Reserved for future application.
46	**reverse-phase, or phase-balance, current relay** is a device which functions when the polyphase currents are of reverse-phase sequence, or when the polyphase currents are unbalanced or contain negative phase-sequence components above a given amount.
47	**phase-sequence voltage relay** is a device which functions upon a predetermined value of polyphase voltage in the desired phase sequence.
48	**incomplete sequence relay** is a device which returns the equipment to the normal, or off, position and locks it out if the normal starting, operating or stopping sequence is not properly completed within a predetermined time.
49	**machine, or transformer, thermal relay** is a device which functions when the temperature of an a-c machine armature, or of the armature or other load carrying winding or element of a d-c machine, or converter or power rectifier or power transformer (including a power rectifier transformer) exceeds a predetermined value.
50	**instantaneous overcurrent, or rate-of-rise relay** is a device which functions instantaneously on an excessive value of current, or on an excessive rate of current rise, thus indicating a fault in the apparatus or circuit being protected.
51	**a-c time overcurrent relay** is a device with either a definite or inverse time characteristic which functions when the current in an a-c circuit exceeds a predetermined value.
52	**a-c circuit breaker** is a device which is used to close and interrupt an a-c power circuit under normal conditions or to interrupt this circuit under fault or emergency conditions.

/more/

Table 9-1. Protective Device Numbering and Functions (*Continued*)

DEVICE NUMBER	DEFINITION AND FUNCTION
53	**exciter or d-c generator relay** is a device which forces the d-c machine field excitation to build up during starting or which functions when the machine voltage has built up to a given value.
54	**high-speed d-c circuit breaker** is a circuit breaker which starts to reduce the current in the main circuit in 0.01 second or less, after the occurrence of the d-c overcurrent or the excessive rate of current rise.
55	**power factor relay** is a device which operates when the power factor in an a-c circuit becomes above or below a predetermined value.
56	**field application relay** is a device which automatically controls the application of the field excitation to an a-c motor at some predetermined point in the slip cycle.
57	**short-circuiting or grounding device** is a power or stored energy operated device which functions to short-circuit or to ground a circuit in response to automatic or manual means.
58	**power rectifier misfire relay** is a device which functions if one or more of the power rectifier anodes fails to fire.
59	**overvoltage relay** is a device which functions on a given value of overvoltage.
60	**voltage balance relay** is a device which operates on a given difference in voltage between two circuits.
61	**current balance relay** is a device which operates on a given difference in current input or output of two circuits.
62	**time-delay stopping, or opening, relay** is a time-delay device which serves in conjunction with the device which initiates the shutdown, stopping or opening operation in an automatic sequence.
63	**liquid or gas pressure, level, or flow relay** is a device which operates on given values of liquid or gas pressure, flow or level, or on a given rate of change of these values.

/more/

Table 9-1. Protective Device Numbering and Functions (*Continued*)

DEVICE NUMBER	DEFINITION AND FUNCTION
64	**ground protective relay** is a device which functions on failure of the insulation of a machine, transformer or of other apparatus to ground, or on flashover of a d-c machine to ground. **note:** This function is assigned only to a relay which detects the flow of current from the frame of a machine or enclosing case or structure of a piece of apparatus to ground, or detects a ground on a normally ungrounded winding or circuit. It is not applied to a device connected in the secondary circuit or secondary neutral of a current transformer, or current transformers, connected in the power circuit of a normally grounded system.
65	**governor** is the equipment which controls the gate or valve opening of a prime mover.
66	**notching, or jogging, device** functions to allow only a specified number of operations of a given device, or equipment, or a specified number of successive operations within a given time of each other. It also functions to energize a circuit periodically, or which is used to permit intermittent acceleration or jogging of a machine at low speeds for mechanical positioning.
67	**a-c directional overcurrent relay** is a device which functions on a desired value of a-c overcurrent flowing in a predetermined direction.
68	**blocking relay** is a device which initiates a pilot signal for blocking of tripping on external faults in a transmission line or in other apparatus under predetermined conditions, or co-operates with other devices to block tripping or to block reclosing on an out-of-step condition or on power swings.
69	**permissive control device** is generally a two-position, manually operated switch which in one position permits the closing of a circuit breaker, or the placing of an equipment into operation, and in the other position prevents the circuit breaker or the equipment from being operated.
70	**electrically operated rheostat** is a rheostat which is used to vary the resistance of a circuit in response to some means of electrical control.
71	Reserved for future application. *|more|*

Table 9-1. Protective Device Numbering and Functions (*Continued*)

DEVICE NUMBER	DEFINITION AND FUNCTION
72	**d-c circuit breaker** is used to close and interrupt a d-c power circuit under normal conditions or to interrupt this circuit under fault or emergency conditions.
73	**load-resistor contactor** is used to shunt or insert a step of load limiting, shifting, or indicating resistance in a power circuit, or to switch a space heater in circuit, or to switch a light, or regenerative, load resistor of a power rectifier or other machine in and out of circuit.
74	**alarm relay** is a device other than an annunciator, as covered under device No. 30, which is used to operate, or to operate in connection with, a visual or audible alarm.
75	**position changing mechanism** is the mechanism which is used for moving a removable circuit breaker unit to and from the connected, disconnected, and test positions.
76	**d-c overcurrent relay** is a device which functions when the current in a d-c circuit exceeds a given value.
77	**pulse transmitter** is used to generate and transmit pulses over a telemetering or pilot-wire circuit to the remote indicating or receiving device.
78	**phase angle measuring, or out-of-step protective relay** is a device which functions at a predetermined phase angle between two voltages or between two currents or between voltage and current.
79	**a-c reclosing relay** is a device which controls the automatic reclosing and locking out of an a-c circuit interrupter.
80	Reserved for future application.
81	**frequency relay** is a device which functions on a predetermined value of frequency—either under or over or on normal system frequency—or rate of change of frequency.
82	**d-c reclosing relay** is a device which controls the automatic closing and reclosing of a d-c circuit interrupter, generally in response to load circuit conditions.
83	**automatic selective control, or transfer, relay** is a device which operates to select automatically between /more/

Table 9-1. Protective Device Numbering and Functions (*Continued*)

DEVICE NUMBER	DEFINITION AND FUNCTION
83 (con't.)	certain sources or conditions in an equipment, or performs a transfer operation automatically.
84	**operating mechanism** is the complete electrical mechanism or servo-mechanism, including the operating motor, solenoids, position switches, etc., for a tap changer, induction regulator or any piece of apparatus which has no device function number.
85	**carrier, or pilot-wire, receiver relay** is a device which is operated or restrained by a signal used in connection with carrier-current or d-c pilot-wire fault directional relaying.
86	**locking-out relay** is an electrically operated hand or electrically reset device which functions to shut down and hold an equipment out of service on the occurrence of abnormal conditions.
87	**differential protective relay** is a protective device which functions on a percentage or phase angle or other quantitative difference of two currents or of some other electrical quantities.
88	**auxiliary motor, or motor generator** is one used for operating auxiliary equipment such as pumps, blowers, exciters, rotating magnetic amplifiers, etc.
89	**line switch** is used as a disconnecting or isolating switch in an a-c or d-c power circuit, when this device is electrically operated or has electrical accessories, such as an auxiliary switch, magnetic lock, etc.
90	**regulating device** functions to regulate a quantity, or quantities, such as voltage, current, power, speed, frequency, temperature, and load, at a certain value or between certain limits for machines, tie lines or other apparatus.
91	**voltage directional relay** is a device which operates when the voltage across an open circuit breaker or contactor exceeds a given value in a given direction.
92	**voltage and power directional relay** is a device which permits or causes the connection of two circuits when the voltage difference between them exceeds a given value in a predetermined direction and causes these

/more/

Table 9-1. Protective Device Numbering and Functions (*Concluded*)

DEVICE NUMBER	DEFINITION AND FUNCTION
92 (con't.)	two circuits to be disconnected from each other when the power flowing between them exceeds a given value in the opposite direction.
93	**field changing contactor** functions to increase or decrease in one step the value of field excitation on a machine.
94	**tripping, or trip-free, relay** is a device which functions to trip a circuit breaker, contactor, or equipment, or to permit immediate tripping by other devices; or to prevent immediate reclosure of a circuit interrupter, in case it should open automatically even though its closing circuit is maintained closed.
95	Reluctance torque synchrocheck
96	Autoloading relay
97 98 99	Used only for specific applications on individual installations where none of the assigned numbered functions from 1 to 94 is suitable.

notes:	[1] A similar series of numbers, starting with 201 instead of 1, shall be used for those device functions in a machine, feeder or other equipment when these are controlled directly from the supervisory system. Typical examples of such device functions are 201, 205, and 294.
	[2] A suffix X, Y, or Z denotes an auxiliary relay.
	[3] TC refers to trip coil.
	[4] CS refers to control switch.
	[5] N, G refers to neutral and ground respectively.

A common way in which power is supplied to a breaker for tripping purposes is through a doc source using 125- or 250-volt station battery.

Figure 9-3 shows a typical electromechanical overcurrent relay schematic. Relay 51X is used as the seal-in relay, which holds in the circuit until it is reset. For a three-phase circuit, three of these relays are required.

Figure 9-4 illustrates a typical tripping circuit. Notice that protective relay contacts are connected in parallel so that any one will trip the break-

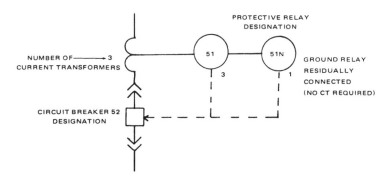

Figure 9-1. One-Line Diagram Showing Protective Relays

er under fault conditions. The control switch (CS) can be manually used to trip the breaker. In order to simplify the protective relay elementary, many times a single contact is used to represent the three. Figure 9-5 presents a simplified protective relay elementary. To close the breaker, frequently only a manual close switch is used. A typical elementary used to close the circuit breaker is illustrated in Figure 9-6.

The Instantaneous Overcurrent Relay (Device 50)
The instantaneous relay is usually combined with the overcurrent relay. The instantaneous attachment can be used, or it can be disconnected from service if it is not required. Figure 9-7 illustrates a typical one-line diagram and schematic utilizing the 50, 51 and 51N protective relays.

THE GROUND OVERCURRENT RELAY (Devices 50N and 50G)

Ground-fault protection has been required since the 1971 NEC. Ground-fault protection saves lives by minimizing damage to circuit conductors and other equipment and safeguarding persons who may simultaneously make contact with electrical equipment and a low resistance path to ground. Ground overcurrent protection can be provided either by overcurrent or instantaneous overcurrent relays. There are three common connections used for ground overcurrent relays, namely residual connection, ground sensor connection, and the neutral CT connection.

Figure 9-8 illustrates the residual grounding scheme. Notice that the one-line diagram of Figure 9-7 has been detailed to show the three-phase

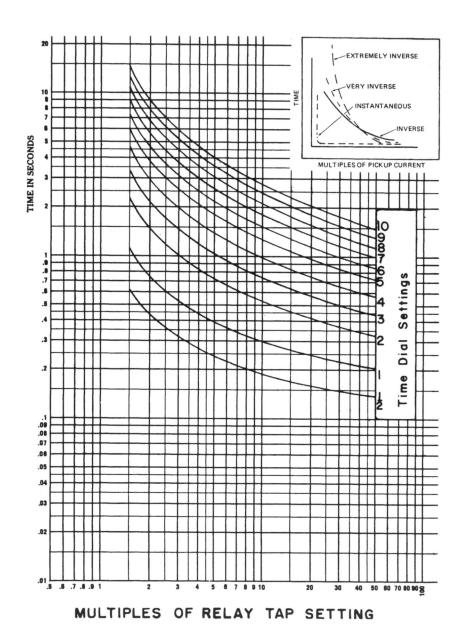

MULTIPLES OF RELAY TAP SETTING

Figure 9-2. Characteristics of Overcurrent Relays

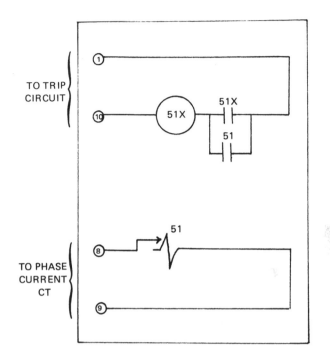

Figure 9-3. Typical Protective Relay (51) Schematic

connections. The residual relaying scheme detects ground-fault current by measuring the current remaining in the secondary of the three-phase of the circuit as transformed by the current transformers. Care must be taken to set the pick-up of the relay above the level anticipated by unbalanced single-phase loads. Due to the possible unbalances caused by unequal current transformer saturation on phase faults and transformer energizing inrush currents, the instantaneous overcurrent relay is seldom used.

The ground sensing scheme is illustrated in Figure 9-9. This scheme uses zero sequence current transformers to detect on ground faults the unbalances in the magnetic flux surrounding the three-phase conductors. Zero sequence current transformers detect when the vectorial summation of the currents is not zero.

The instantaneous or overcurrent relay can be used with this scheme. The installation of the zero-sequence window current transformer should not enclose the equipment ground conductor or the conductor shielding. With the ground sensing scheme it is possible to detect and clear system faults as small as 15 amperes.

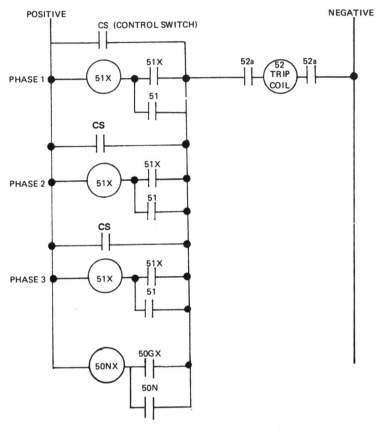

Figure 9-4. Typical Trip Circuit of Protective Relays

The neutral grounding scheme illustrated in Figure 9-10 is used commonly with resistively grounded transformers. In this scheme the ground-fault current is sensed by the current transformer in the resistively grounded neutral conductor.

PARTIAL DIFFERENTIAL OR SUMMATION RELAYING

This protective relaying scheme is commonly used to detect and isolate faults without affecting other portions of the system. Figure 9-11 illustrates a typical partial differential relaying scheme. In this scheme the tie breakers are nominally closed. If a fault occurs on BUS G, breakers C and D should trip, leaving breaker A unaffected. Likewise a fault on BUS

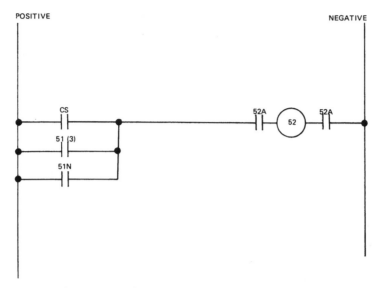

Figure 9-5. Simplified Protective Relay Elementary

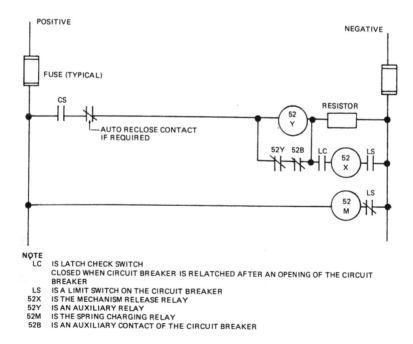

NOTE
LC   IS LATCH CHECK SWITCH
CLOSED WHEN CIRCUIT BREAKER IS RELATCHED AFTER AN OPENING OF THE CIRCUIT BREAKER
LS   IS A LIMIT SWITCH ON THE CIRCUIT BREAKER
52X   IS THE MECHANISM RELEASE RELAY
52Y   IS AN AUXILIARY RELAY
52M   IS THE SPRING CHARGING RELAY
52B   IS AN AUXILIARY CONTACT OF THE CIRCUIT BREAKER

Figure 9-6. Typical Close Elementary of a Circuit Breaker

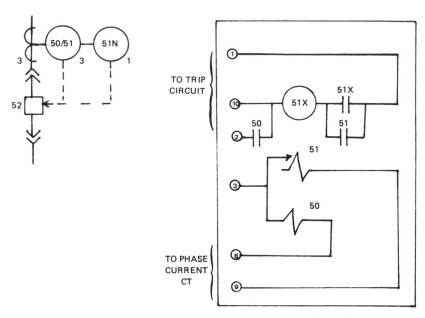

Figure 9-7. Typical One-line and Protective Relay (50/51) Schematic

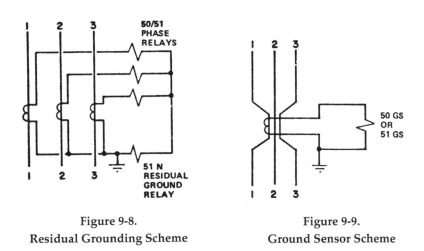

Figure 9-8.	Figure 9-9.
Residual Grounding Scheme	**Ground Sensor Scheme**

F should trip breakers A and C and leave breaker D unaffected. One way of accomplishing this is to connect the current transformers of protective relays such that they will only pick up when the fault currents through the pair of current transformers flow in opposite directions. For example, a fault on BUS F will cause fault currents flowing to the fault. The current

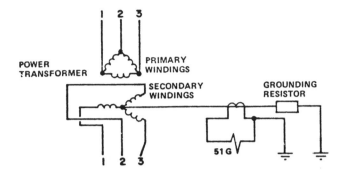

Figure 9-10. Neutral Grounding Relay Scheme

Table 9-1. Maximum Rating or Setting of Motor Branch-circuit, Short-circuit, and Ground-fault Protective Devices.

	Percentage of Full-Load Current			
Type of Motor	Nontime Delay Fuse[1]	Dual Element (Time-Delay) Fuse[1]	Instantaneous Trip Breaker	Inverse Time Breaker[2]
Single-phase motors	300	175	800	250
AC polyphase motors other than wound-rotor	300	175	800	250
Squirrel cage — other than Design B energy-efficient	300	175	800	250
Design B energy-efficient	300	175	1100	250
Synchronous[3]	300	175	800	250
Wound rotor	150	150	800	150
Direct current (constant voltage)	150	150	250	150

transformers associated with protective relays for breaker A will sense currents in opposite directions, thus activating these relays. On the other hand, a fault on BUS G will cause currents to flow in the same direction through the current transformers; thus these relays will not operate under this condition.

DIFFERENTIAL PROTECTIVE RELAY (Device 87)

The differential protective relay is used for protecting AC rotating machinery, generators and transformers. This relay operates on the difference between two currents. A typical application is illustrated in Figure 9-12. This figure shows the differential principle applied to a single-phase winding of electrical equipment, such as a generator. In this application a current balance relay is used to provide what is called "percentage differ-

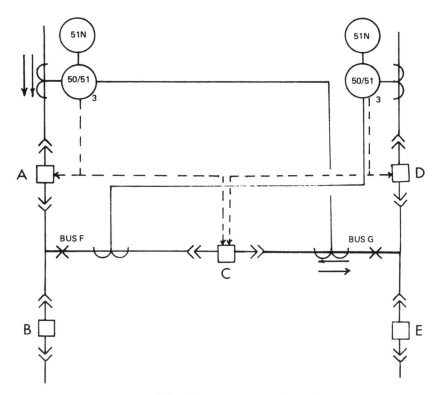

Figure 9-11. Partial Differential Protective Relay Scheme

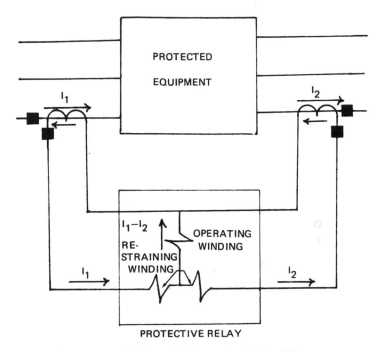

Figure 9-12. Differential Protective Relay Scheme

ential" relaying. The current transformers are connected to the equipment to be protected.

The current from each transformer flows through a restraining coil. The purpose of the restraining coil is to prevent undesired relay operation as a result of a mismatch in current transformers. When a fault does occur, the operating relay sees a percentage increase in current, and the relay operates.

Notice on Figure 9-12 the introduction of polarity marks on the current transformers. Polarity identification marks are as follows:

• Current flows into the polarity mark for primary connections. Current flows out of the polarity mark for secondary connections.

• Voltage drops from polarity to nonpolarity for primary and secondary connections.

See Figure 9-13 for an illustration of polarity marks.

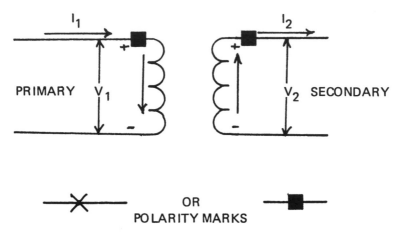

Figure 9-13. Polarity and Circuit Diagram

UNDERVOLTAGE RELAY (Device 27) AND
OVERVOLTAGE RELAY (Device 59)

The undervoltage and overvoltage relays are used wherever protection is required for these conditions. These relays usually operate continuously energized and are adjusted to drop out at any voltage within their calibration range. Figure 9-14 illustrates a one-line diagram for undervoltage and overvoltage relays. In this particular scheme, breaker A is tripped by undervoltage relay. An auxiliary contact from breaker A trips breaker B after breaker A is tripped. The overvoltage relay closes breaker B when the preset voltage level is reached. The characteristics of undervoltage relays are indicated in Figure 9-15.

APPLYING SOLID-STATE PROTECTIVE RELAYS

Solid-state protective relays offer significantly improved characteristics over electromechanical relays and are available for the most important areas of the system. Solid-state techniques allow improvement in sensitivity and temperature stability, plus effective transient surge protection.
The main advantages in using solid-state relays are:
• Flexible settings
• Dynamic performance
• Improved instantaneous

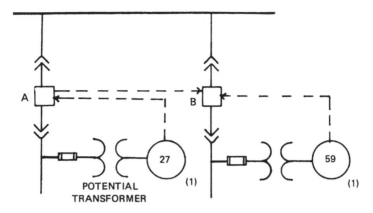

Figure 9-14. One-Line Diagram for Undervoltage and Overvoltage Relay

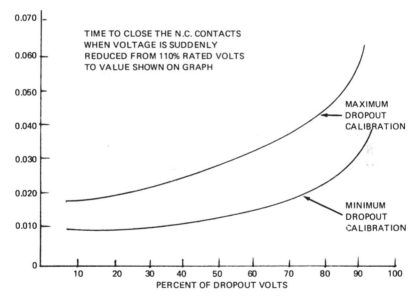

Figure 9-15. Characteristics of Undervoltage Relays

- Very low burden
- Easy testing
- Reduction in panel space
- Improved indication
- Better immunity to shock
- Good repeatability

These relays are available in time overcurrent relays, instantaneous overcurrent relays, voltage relays, directional relays, reclosing relays, ground-fault relays, direct-current relays and timing relays.

Some of the advantages of solid-state relays are summarized below.

Easy Testing and Flexible Settings

Installation testing is performed by depressing test buttons on the relay; thus test equipment is not required. These test buttons allow for initial settings, operational and wiring checks, and maintenance tests. A rough timing check can be performed with the second hand of an ordinary wristwatch. More precise tests can also be made.

Dynamic Performance

The solid-state relay can be thought of as a fine-tuned electrome-chanical protective device. The conventional electromechanical overcur-rent relay is the induction disc type. The induction disc element, which is either copper or aluminum, rotates between the pole faces of an electro-magnet. There are two methods commonly used to rotate the induction disc. The shaded pole method is illustrated in Figure 9-16. In this method a portion of the electromagnet pole face is short-circuited by a copper ring or coil to cause the flux in the shaded portion to lag the flux in the un-shaded portion.

The second method is referred to as the wattmetric type and uses one set of coils above and below the disc. In both methods the moving contact

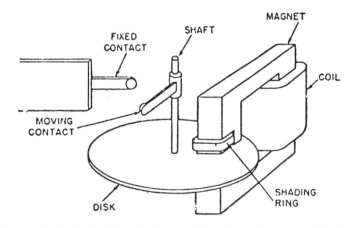

Figure 9-16. Shaded Pole Induction Disc Type Overcurrent Relay

is carried on the rotating shaft. In the induction-disc type of overcurrent relay, the disc continues to rotate after the starting current has decreased to a low value. This overtravel means that, to avoid nuisance tripping, the time dial setting of the relay is set several positions higher than is desirable. Solid-state relays do not have rotating parts; thus the problem of overtravel does not occur. This means that the characteristics of the relay can be adjusted to what is required, without allowances for dynamic effects such as overtravel.

The solid-state overcurrent relay consists of printed circuit boards that produce a DC output voltage when the input AC current exceeds a given value. Each overcurrent function usually consists of an input transformer, overcurrent module, and a resistor-zener-diode protective network. The overcurrent module consists of a setting circuit, phase-splitter circuit, sensing circuit, amplifier circuit, feedback circuit and an output circuit. The overcurrent module can either be a single-input module with one output, or a dual-input module with a single output.

Low Burden

Low-volt ampere (burden) requirements for protective relays mean that more relays may be connected in series, the lowest tap of a multi-ratio CT can be used, and the auxiliary CT can be stepped up for residual current sensitivity. This translates into dollar and space savings as a result of low-volt ampere requirements on instrument transformers. Another savings is that bush-mounted current transformers are not required with solid-state relays. Solid-state relays are compatible with bushing-mounted current transformers in low- and medium-voltage switchgear applications.

Reduction in Panel Space

A comparison of solid-state relay sizes with their electromechanical counterparts indicates a space savings of at least one third. For complex relay schemes space savings can be 75 percent or more.

Improved Indication

The solid-state protective relay has a target which operates independently of the trip-coil current and depends only on the proper functioning of the relay. This helps in troubleshooting in the event of a broken trip-coil connection. With the new design it is possible to have an indicator without a seal-in contact in parallel with the relay's measuring contacts. Often this results in simplification of complex schemes.

Better Immunity to Shock

Since the solid-state relay has no moving parts, it is in many cases better suited for earthquake-prone locations. Solid-state relays have been tested and have withstood accelerations of up to 109 and higher.

Reliability

Solid-state devices are reliable, have good repeatability, and are economical for industrial applications. The major protective relay manufacturers now offer solid-state relays as part of their line. Over the last few years solid-state protective devices have proved to be a new tool for system protection.

SUMMARY FOR SOLID STATE RELAYS

"A solid state relay is a control relay with an isolated input and output, whose functions are achieved by means of electronic components without the use of moving parts such as those found in electromechanical relays. These solid state electrical relays are ideal for applications that have many contact closures, since solid state relay switches offer a greatly extended life compared to electromechanical relays."

"Solid state relays (SSR) are similar to electromechanical relays, in that both use a control circuit and a separate circuit for switching the load. When voltage is applied to the input of the SSR, the relay switch is energized by a light emitting diode. The light from the diode is beamed into a light sensitive semiconductor which, in the case of zero voltage crossover control relays, conditions the control circuit to turn on the output of the solid state switch at the next zero voltage crossover."

"In the case of non-zero voltage crossover relays, the output of the solid state switch is turned on at the precise voltage occurring at the time. Removal of the input power disables the control circuit, and the solid state relay switch is turned off when the load current passes through the zero point of its cycle."

Solid state relays have features which electromechanical relays do not, such as:

- Long life
- Shock and vibration resistance
- Generation of RFI, EMI

- No contact bounce
- Arcless switching
- Acoustic noise
- Zero voltage switching
- IC compatibility
- Immunity to humidity, salt spray and dirt"*

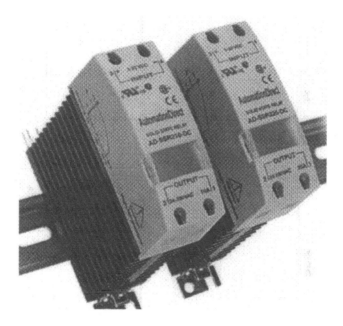

Figure 9-17. Solid-state Relays (*Source: Automation Direct*)

---

*Source: automation direct catalog

Chapter 10

Energy
Economic Analysis

To justify the energy investment cost, a knowledge of life-cycle costing is required.

The life-cycle cost analysis evaluates the total owning and operating cost. It takes into account the time value of money and can incorporate fuel cost escalation into the economic model. This approach is also used to evaluate competitive projects. In other words, the life-cycle cost analysis considers the cost over the life of the system rather than just the first cost.

THE TIME VALUE OF MONEY CONCEPT

To compare energy utilization alternatives, it is necessary to convert all cash flow for each measure to an equivalent base. The life-cycle cost analysis takes into account the time value of money; thus a dollar in hand today is more valuable than one received at some time in the future. This is why a time value must be placed on all cash flows into and out of the company.

DEVELOPING CASH FLOW MODELS

The cash flow model assumes that cash flows occur at discrete points in time as lump sums and that interest is computed and payable at discrete points in time.

To develop a cash flow model that illustrates the effect of "compounding" of interest payments, the cash flow model is developed as follows:

End of Year 1: $P + i(P) = (1 + i)P$

Year 2: $(1 + i) P + (1 + i) Pi = (1 + i)P [(1 + i)]$

         $= (1 + i)^2 P$

Year 3: $(1 + i)^3 P$

Year n $(1 + i)^n P$ or $F = (1 + i)^n P$

Where P = present sum

      I = interest rate earned at the end of each interest period

      n = number of interest periods

      F = future value

$(1 + i)^n$ is referred to as the "Single Payment Compound Amount" factor (F/P) and is tabulated for various values of i and n in Appendix A.

The cash flow model can also be used to find the present value of a future sum F.

$$P = \left(\frac{1}{(1 + i)^n} \right) \times F$$

Cash flow models can be developed for a variety of other types of cash now as illustrated in Figure 10-1.

To develop the cash flow model for the "Uniform Series Compound Amount" factor, the following cash flow diagram is drawn.

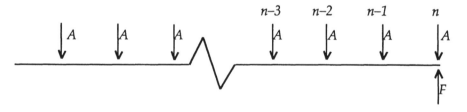

Where A is a uniform series of year-end payments and F is the future sum of A payments for n interest periods.

The A dollars deposited at the end of the nth period earn no interest and, therefore, contribute A dollars to the fund. The A dollars deposited at the end of the $(n - 1)$ period earn interest for 1 year and will, therefore, contribute $A (1 + i)$ dollars to the fund. The A dollars deposited at the end of the $(n - 2)$ period earn interest for 2 years and will, therefore, contribute

$A(1 + i)^2$. These years of earned interest in the contributions will continue to increase in this manner, and the A deposited at the end of the first period will have earned interest for $(n - 1)$ periods. The total in the fund F is, thus, equal to $A + A(1 +i) + A(1 +i)^2 + A(1 +i)^3 + A(1 +i)^4 +...+ A(1 +i)^{n-2} + A(1 +i)^{n-1}$. Factoring out A,

(1) $F = A[1 + (1 +i) + (1 +i)^2 ... +(1 +i)^{n-2} + (1 +i)^{n-1}$
      Multiplying both sides of this equation by $(1 +i)$;

(2) $(1+i)F = A[(1+i)+(1+i)^2 + (1+i)^3 + ... + (1+i)^{n-1} + (1+i)n]$
      Subtracting equation (1) from (2):

$$(1+i)F - F = A[(1+i) + (1+i)^2 + (1+i)^3$$
$$+ (1+i)^{n-1} +(1 +i)^n] - A[1 + (1 +i)$$
$$+(1+i)^2 + ... + (1+i)^{n-2} + (1+i)^{n-1}]$$
$$iF = A[(1+i)^{n-1}]$$

$$F = A\left[\frac{(1 + i)^{n-1}}{i}\right]$$

Interest factors are seldom calculated. They can be determined from computer programs and interest tables included in the Appendix. Each factor is defined when the number of periods (n) and interest rate (i) are specified. In the case of the gradient present worth factor the escalation rate must also be stated.

The three most commonly used methods in life-cycle costing are the annual cost, present worth, and rate-of-return analysis.

In the present worth method a minimum rate of return (i) is stipulated. All future expenditures are converted to present values using the interest factors. The alternative with lowest effective first cost is the most desirable.

A similar procedure is implemented in the annual cost method. The difference is that the first cost is converted to an annual expenditure. The alternative with the lowest effective annual cost is the most desirable.

In the rate-of-return method, a trial-and-error procedure is usually required. Interpolation from the interest tables can determine what rate of return (i) will give an interest factor that will make the overall cash flow balance. The rate-of-return analysis gives a good indication of the overall ranking of independent alternates.

Single Payment Compound Amount - *F/P*

The *F/P* factor is the future value of one dollar in "*n*" periods at interest of "*i*" percent.

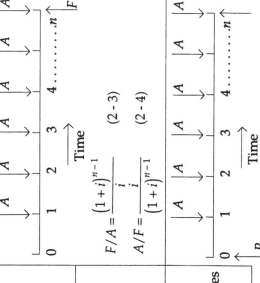

$$F/P = (1 + i)^n \qquad (2\text{-}1)$$

Single Payment Present Worth - *P/F*

The *P/F* factor is the present worth of one dollar, in "*n*" periods from now at interest of "*i*" percent.

$$P/F = \frac{1}{(1+i)^n} \qquad (2\text{-}2)$$

Uniform Series Compound Amount - *F/A*

The *F/A* factor is the future value of a uniform series of one dollar deposits.

$$F/A = \frac{(1+i)^{n-1}}{i} \qquad (2\text{-}3)$$

Sinking Fund Payment - *A/F*

The *A/F* factor is the uniform series of deposits whose future value is one dollar.

$$A/F = \frac{i}{(1+i)^{n-1}} \qquad (2\text{-}4)$$

Uniform Series Present Worth - *P/A*

The *P/A* factor is the present value of uniform series of one dollar deposits.

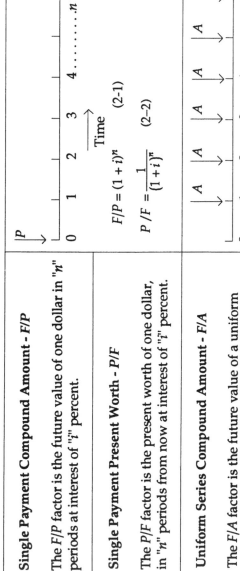

Capital Recovery - A/P

The *A/P* factor is the uniform series of deposits whose present value is one dollar.

$$P/A = \frac{(1+i)^n - 1}{i(1+i)^n} \qquad (2\text{-}5)$$

$$A/P = \frac{i(1+i)^n}{(1+i)^n - 1} \qquad (2\text{-}6)$$

Gradient Present Worth - GPW

The GPW factor is the present value of a gradient series.

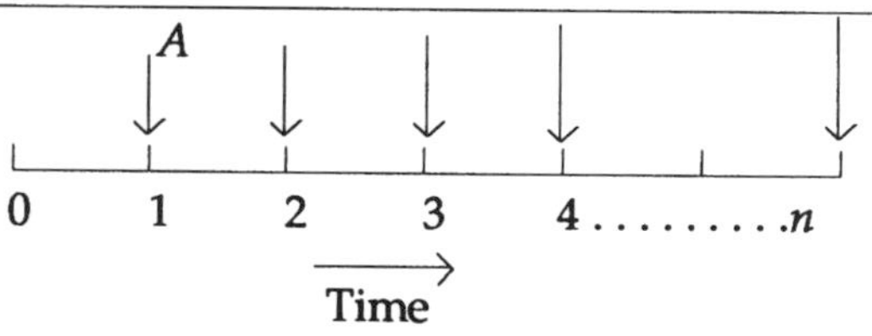

$$GPW = P/A = \frac{\frac{1+e}{1+i}\left[1 - \left(\frac{1+e}{1+i}\right)^n\right]}{1 - \frac{1+e}{1+i}} \qquad (2\text{-}7)$$

NOTES

where

P is the present worth (occurs at the beginning of the interest period).

F is the future worth (occurs at the end).

n is the number of periods that the interest is compounded.

i is the interest rate or desired rate of return.

A is the uniform series of deposits (occurs at the end of the interest period).

e is the escalation rate

Figure 10-1. Interest Factors

The effect of escalation in fuel costs can influence greatly the final decision. When an annual cost grows at a steady rate, it may be treated as a gradient and the gradient present worth factor can be used.

Special appreciation is given to Rudolph R. Vaneck and Dr. Robert Brown for use of their specially designed interest and escalation tables used in this text.

When life-cycle costing is used to compare several alternatives, the differences between costs are important. For example, if one alternate forces additional maintenance or an operating expense to occur, then these factors, as well as energy costs, need to be included. Remember, what was previously spent for the item to be replaced is irrelevant. The only factor to be considered is whether the new cost can be justified based on projected savings over its useful life.

PAYBACK ANALYSIS

The simple payback analysis is sometimes used instead of the methods previously outlined. The simple payback is defined as initial investment divided by annual savings after taxes. The simple payback method does not take into account the effect of interest or escalation rate.

Since the payback period is relatively simple to calculate, and due to the fact managers wish to recover their investment as rapidly as possible, the payback method is frequently used.

It should be used in conjunction with other decision-making tools. When used by itself as the principal criterion, it may result in choosing less profitable investments that yield high initial returns for short periods, as compared with more profitable investments which provide profits over longer periods of time.

Example SIM 10-1
An electrical energy audit indicates electrical motor consumption is 4×10^6 kWh per year. By upgrading the motor spares with high efficiency motors, a 10% savings can be realized. The additional cost for these motors is estimated at $80,000. Assuming an 8¢ per kWh energy charge and 20-year life, is the expenditure justified based on a minimum rate of return of 20% before taxes? Solve the problem using the present worth, annual cost, and rate-of-return methods.

Analysis

Present Worth Method

	Alternate 1 *Present Method*	Alternate 2 *Use High Efficiency* *Motor Spares*
(1) First Cost (P)	—	$80,000
(2) Annual Cost (A)	$4 \times 10^6 \times .08$ = $320,000	$.9 \times $320,000$ = $288,000
USPW (Table A-4)	4.87	4.87
(3) A × 4.87 =	$1,558,400	$1,402,560
Present Worth	$1,558,400	$1,482,560
(1) + (3)		Choose Alternate with Lowest Present Worth Cost

Annual Cost Method

	Alternate 1	Alternate 2
(1) First Cost (P)	—	$80,000
(2) Annual Cost (A)	$320,000	$288,000
CR (Table A-4)	.2	.2
(3) P × .2	—	$16,000
Annual Cost	$320,000	$304,000
(2) + (3)		Choose Alternate with Lowest Annual Cost

Rate of Return Method

$$P = (\$320,000 - \$288,000)$$

$$P/A = \frac{80,000}{32,000} = 2.5$$

What value of i will make $P/A = 2.5$? $i = 40\%$ (From Table A-7).

Example SIM 10-2

Show the effect of 10% escalation on the rate-of-return analysis given the

Energy equipment investment	=	$20,000
After tax savings	=	$ 2,600
Equipment life (n)	=	15 years

Analysis

Without escalation

$$CR = \frac{A}{P} = \frac{2,600}{20,000} = .13$$

From Table A-1, the rate of return in 10%.
With 10% escalation assumed:

$$GPW = \frac{P}{A} = \frac{20,000}{2,600} = 7.69$$

From Table A-11, the rate of return is 21%.

Thus we see that taking into account a modest escalation rate can dramatically affect the justification of the project.

TAX CONSIDERATIONS

Depreciation

Depreciation affects the "accounting procedure" for determining profits and losses and the income tax of a company. In other words, for tax purposes the expenditure for an asset such as a pump or motor cannot be fully expensed in its first year. The original investment must be charged off for tax purposes over the useful life of the asset. A company wishes to expense an item as quickly as possible.

The Internal Revenue Service allows several methods for determining the annual depreciation rate.

Straight-line Depreciation: The simplest method is referred to as a straight-line depreciation and is defined as

$$D = \frac{P-L}{n} \hspace{3cm} \textit{Formula (10-8)}$$

Where
  D is the annual depreciation rate
  L is the value of equipment at the end of its useful life, commonly referred to as salvage value

n is the life of the equipment which is determined by Internal Revenue Service guidelines

P is the initial expenditure.

Sum-of-years Digits: Another method is referred to as the sum-of-years digits. In this method the depreciation rate is determined by finding the sum of digits using the following formula:

$$N = n \; \frac{n+1}{2} \qquad \qquad Formula\ (10\text{-}9)$$

Where n is the life of equipment.

Each year's depreciation rate is determined as follows:

First year $\qquad D = \dfrac{n}{N} \; (P - L) \qquad \qquad Formula\ (10\text{-}10)$

Second year $\qquad D = \dfrac{n-1}{N} \; (P - L) \qquad \qquad Formula\ (10\text{-}11)$

n year $\qquad D = \dfrac{1}{N} \; (P - L) \qquad \qquad Formula\ (10\text{-}12)$

Declining-balance Depreciation: The declining-balance method allows for larger depreciation charges in the early years, which is sometimes referred to as fast write-off.

The rate is calculated by taking a constant percentage of the declining undepreciated balance. The most common method used to calculate the declining balance is to predetermine the depreciation rate. Under certain circumstances a rate equal to 200% of the straight-line depreciation rate may be used. Under other circumstances the rate is limited to 1-1/2 or 1-1/4 times as great as straight-line depreciation. In this method the salvage value or undepreciated book value is established once the depreciation rate is pre-established.

To calculate the undepreciated book value, Formula 10-13 is used:

$$D = 1 - \left(\frac{L}{P}\right)^{1/N} \qquad \qquad Formula\ (10\text{-}13)$$

Where
    D is the annual depreciation rate
    L is the salvage value
    P is the first cost

Example SIM 10-3

Calculate the depreciation rate using the straight-line, sum-of-years digit, and declining-balance methods.

Salvage value is 0.

$n = 5$ years

$P = 150,000$

For declining balance use a 200% rate.

Straight-line Method

$$D = \frac{P - L}{n} = \frac{150,000}{5} = \$30,000 \text{ per year}$$

Sum-of-years Digits

$$N = \frac{n(n + 1)}{2} = \frac{5(6)}{2} = 15$$

$$D_1 = \frac{n}{N} (P) = \frac{5}{15} (150,000) = 50,000$$

N	P
1 =	$54,000
2 =	40,000
3 =	30,000
4 =	20,000
5 =	10,000

Declining-balance Method

$D = 2 \times 20\% = 40\%$ (Straight-line Depreciation Rate = 20%)

Year	Undepreciated Balance At Beginning of Year	Depreciation Charge
1	150,000	60,000
2	90,000	36,000
3	54,000	21,600
4	32,400	12,960
5	19,440	7,776
	TOTAL	138,336

Undepreciated Book Value (150,000 – 138,336) = $11,664

Cogeneration Equipment Depreciation

Most cogeneration equipment is depreciated over a 15- or 20-year period, depending on the particular type of equipment involved, using the 150% declining balance method switching to straight-line to maximize deductions. Gas and combustion turbine equipment used to produce electricity for sale is depreciated over a 15-year period. Equipment used in the steam power production of electricity for sale (including combustion turbines operated in combined cycle with steam units), as well as assets used to produce steam for sale, are normally depreciated over a 20-year period.

However, most electric and steam generation equipment owned by a taxpayer and producing electric or thermal energy for use by the taxpayer in its industrial process and plant activity, and not ordinarily for sale to others, is depreciated over a 15-year period. Electrical and steam transmission and distribution equipment will be depreciated over a 20-year period at the same 150 percent declining balance rate.

Energy Efficiency Equipment and Real Property Depreciation

Energy conservation equipment, still classified as real property, is depreciated on a straight line basis over a recovery period. Equipment installed in connection with residential real property qualifies for a 27-1/2-year period, while equipment placed in nonresidential facilities is subject to a 31-1/2-year period. Other real property assets are depreciated over the above period, depending on their residential or nonresidential character.

After-tax Analysis

Tax-deductible expenses such as maintenance, energy, operating costs, insurance and property taxes reduce the income subject to taxes.

For the after-tax life-cycle cost analysis and payback analysis, the actual incurred annual savings is given as follows:

$$AS=(1 - I)\, E + ID \qquad\qquad \textit{Formula (10-14)}$$

Where

AS = yearly annual after-tax savings (excluding effect of tax credit)

E = yearly annual energy savings (difference between original expenses and expenses after modification)

D = annual depreciation rate

I = income tax bracket

Formula 10-14 takes into account that the yearly annual energy savings is partially offset by additional taxes that must be paid due to reduced operating expenses. On the other hand, the depreciation allowance reduces taxes directly.

To compute a rate of return which accounts for taxes, depreciation, escalation and tax credits, a cash-flow analysis is usually required. This method analyzes all transactions, including first and operating costs. To determine the after-tax rate of return, a trial and error or computer analysis is required.

The present worth factors tables in the Appendix, can be used for this analysis. All money is converted to the present, assuming an interest rate. The summation of all present dollars should equal zero when the correct interest rate is selected, as illustrated in Figure 10-2.

This analysis can be made assuming a fuel escalation rate by using the gradient present worth interest of the present worth factor.

Example SIM 10-4

Comment on the after-tax rate of return for the installation of a heat-recovery system, given the following:

- First Cost $100,000
- Year Savings   36,363
- Straight-line depreciation life and equipment life of 5 years
- Income tax bracket 34%

Year	1 Investment	2 Tax Credit	3 After Tax Savings (AS)	4 Single Payment Present Worth Factor	(2 + 3) x 4 Present Worth
0	–P				–P
1		+TC	AS_1	$SPPW_1$	$+P_1$
2			AS_2	$SPPW_2$	P_2
3			AS_3	$SPPW_3$	P_3
4			AS_4	$SPPW_4$	P_4
Total					ΣP

$$AS = (1 - I)\,E + ID$$
Trial & Error Solution:
Correct i when $\Sigma P = 0$

Figure 10-2. Cash Flow Rate of Return Analysis

Analysis

$D = 100{,}000/5 = 20{,}000$

$AS = (1 - I)\,E + ID = .66(36{,}363) + .34(20{,}000) = \$30{,}800$

First Trial $i = 20\%$

Investment	After Tax Savings	SPPW 20%	PW
0-100,000			–100,000
1	30,800	.833	25,656
2	30,800	.694	21,374
3	30,800	.578	17,802
4	30,800	.482	14,845
5	30,800	.401	12,350
			$\Sigma - 7{,}972$

Since summation is negative a higher present worth factor is required. Next try is 15%.

Investment	After Tax Savings	SPPW 15%	PW
0-100,000			-100,000
1	30,800	.869	+ 26,765
2	30,800	.756	+ 23,284
3	30,800	.657	+ 20,235
4	30,800	.571	+ 17,586
5	30,800	.497	+ 15,307
			+ 3,177

Since rate of return is bracketed, linear interpolation will be used.

$$\frac{3177 + 7972}{-5} = \frac{3177 - 0}{15 - i\%}$$

$$i = \frac{3177}{2229.6} + 15 = 16.4\%$$

Impact of Fuel Inflation on Life-cycle Costing

As illustrated by SIM 10-2, a modest estimate of fuel inflation has a major impact on improving the rate of return on investment of the project. The problem facing the energy engineer is how to forecast what the future of energy costs will be. All too often no fuel inflation is considered because of the difficulty of projecting the future. In making projections the following guidelines may be helpful:

- Is there a rate increase that can be forecast based on new nuclear generating capacity? In locations such as Georgia, California, and Arizona, electric rates will rise at a faster rate due to commissioning of new nuclear plants and rate increases approved by the Public Service Commission of that state.
- What has been the historical rate increase for the facility? Even with fluctuations there are likely to be trends to follow.
- What events on a national or international level would impact your costs? New state taxes, new production quotas by OPEC, and other factors will affect your fuel prices.
- What do the experts say? Energy economists, forecasting services, and your local utility projections all should be taken into account.

The rate of return on investment becomes more attractive when life-cycle costs are taken into account. Tables A-9 through A-12 can be used to show the impact of fuel inflation on the decision-making process.

Example SIM 10-5
Develop a set of curves that indicate the capital that can be invested to give a rate of return of 15% after taxes for each $1000 saved for the following conditions:
1. The effect of escalation is not considered.
2. A 5% fuel escalation is considered.
3. A 10% fuel escalation is considered.
4. A 14% fuel escalation is considered.
5. A 20% fuel escalation is considered.

Calculate for 5-, 10-, 15-, 20-year life.
Assume straight-line depreciation over useful life, 34% income tax bracket, and no tax credit.

Answer

$$AS = (1 - I)E + ID$$
$$I = 0.34, E = \$1000$$
$$AS = 660 + \frac{0.34P}{N}$$

Thus, the after-tax savings (AS) is composed of two components. The first component is a uniform series of $660 escalating at e percent/year. The second component is a uniform series of $0.34P/N$.

Each component is treated individually and convened to present day values using the GPW factor and the USPW factor, respectively. The sum of these two present worth factors must equal P. In the case of no escalation, the formula is

$$P = 660\, P/A + \frac{0.34P}{N}\, P/A$$

In the case of escalation

$$P = 660\, GPW + \frac{0.34P}{N}\, P/A$$

Since there is only one unknown, the formulas can be readily solved. The results are indicated below.

	N=5 $P	N=10 $P	N= 15 $P	N=20 $P
e = 0	2869	4000	4459	4648
e = 10%	3753	6292	8165	9618
e = 14%	4170	7598	10,676	13,567
e = 20%	4871	10,146	16,353	23,918

Figure 10-3 illustrates the effects of escalation. This figure can be used as a quick way to determine after-tax economics of energy utilization expenditures.

Example SIM 10-6

It is desired to have an after-tax savings of 15%. Comment on the investment that can be justified if it is assumed that the fuel rate escalation should not be considered, the annual energy savings is $2000, and the equipment economic life is 10 years.

Comment on the above, assuming a 10% fuel escalation.

Answer

From Figure 10-3, for each $1000 energy savings, an investment of $3600 is justified (or $8000 for a $2000 savings) for which no fuel increase is accounted.

With a 10% fuel escalation rate on investment of $6300 justified for each $1000 energy savings, $12,600 can be justified for $2000 savings. Thus, a 57% higher expenditure is economically justifiable and will yield the same after-tax rate of return of 15% when a fuel escalation of 10% is considered.

Figure 10-3. Effects of Escalation on Investment Requirements

Note: Maximum investment in order to attain a 15% after-tax rate of return on investment for annual savings of $1000.

Chapter 11

Energy Management Systems

ENERGY MANAGEMENT

The availability of computers at moderate costs and the concern for reducing energy consumption have resulted in the application of computer-based controllers to more than just industrial process applications. These controllers, commonly called energy management systems (EMS), can be used to control virtually all non-process energy using pieces of equipment in buildings and industrial plants. Equipment controlled can include fans, pumps, boilers, chillers and lights. This section will investigate the various types of energy management systems which are available and will illustrate some of the methods used to reduce energy consumption.

BASIC TIME CONTROL

One of the simplest and most effective methods of conserving energy in a building is to operate equipment only when it is needed. If, due to time, occupancy, temperature or other means, it can be determined that a piece of equipment does not need to operate, energy savings can be achieved without affecting occupant comfort by turning off the equipment.

One of the simplest devices to schedule equipment operation is the basic timer. A common application of basic timers is scheduling office building HVAC equipment to operate during business hours Monday through Friday and to be off all other times. As is shown in the following problem, significant savings can be achieved through the correct application of basic timers.

Example SIM 11-1

An office building utilizes two 50 hp supply fans and two 15 hp return fans, all of which operate continuously to condition the building. What are the annual savings that result from installing a timer to operate these fans from 7:00 a.m. to 5:00 p.m., Monday through Friday? Assume an electrical rate of $0.16/kWh and motor efficiency of .8.

Answer

Annual operation before timeclock =

52 weeks × 7 days/week × 24 hours/day = 8736 hours

Annual operation after timeclock =

52 × (5 days/week × 10 hours/day) = 2600 hours

Savings = 130 hp × 0.746 kW/hp × (8736-2600) hours × 1/.8 × $0.16/kWh = $119,000

Although most buildings today utilize some version of timer control, the magnitude of the savings value in this example illustrates the importance of correct timeclock operation and the potential for additional costs if this device should malfunction or be adjusted inaccurately. Note that the above example also ignores heating and cooling savings which would result from the installation of a timer.

PROBLEMS WITH BASIC IMERS

Although the use of timers has resulted in significant energy savings, they are being replaced by energy management systems because of problems that include the following:

- Holidays, when the building is unoccupied, cannot easily be taken into account.

- Power failures require the timeclock to be reset, or it is not synchronized with the building schedule.

- There are a limited number of on and off cycles possible each day.

- It is a time-consuming process to change schedules on multiple timeclocks.

Energy management systems are designed to overcome these problems plus provide increased control of building operations.

ENERGY MANAGEMENT SYSTEMS

Recent advances in digital technology, dramatic decreases in the cost of this technology, and increased energy awareness have resulted in the increased application of computer-based controllers (i.e., energy management systems) in commercial buildings and industrial plants. These devices can control from one to a virtually unlimited number of items of equipment.

By concentrating the control of many items of equipment at a single point, the EMS allows the building operator to tailor building operation to precisely satisfy occupant needs. This ability to maximize energy conservation, while preserving occupant comfort, is the ultimate goal of an energy engineer.

Microprocessor Based EMS

Energy management systems can be placed in one of two broad, and sometimes overlapping, categories referred to as microprocessor-based and computer based.

Microprocessor-based systems can control from 1 to 40 input/out-points and can be linked together for additional loads. Programming is accomplished by a keyboard or hand-held console, and an LED display is used to monitor/review operation of the unit. A battery maintains the programming in the event of power failure. (See Figures 11-2 and 11-3.)

Capabilities of this type of EMS are generally pre-programmed so that operation is relatively straightforward. Programming simply involves entering the appropriate parameters for the desired function (e.g., the point number and the on and off times). Microprocessor-based EMS can have any or all of the following capabilities:

- Scheduling
- Duty Cycling
- Demand Limiting

- Optimal Start
- Monitoring
- Direct Digital Control

Scheduling EMS

Scheduling with an EMS is very much the same as it is with a time-clock. Equipment is started and stopped based on the time of day and the day of week. Unlike a timeclock, however, multiple start/stops can be accomplished very easily and accurately. For example, in a classroom, lights can be turned off during morning and afternoon break periods and during lunch). It should be noted that this single function, if accurately programmed and depending on the type of facility served, can account for the largest energy savings attributable to an EMS.

Additionally, holiday dates can be entered into the EMS a year in advance. When the holiday occurs, regular programming is overridden and equipment can be kept off.

Duty Cycling

Most HVAC fan systems are designed for peak load conditions; consequently these fans are usually moving much more air than is needed. Therefore, they can sometimes be shut down for short periods each hour (typically 15 minutes) without affecting occupant comfort. Turning equipment off for pre-determined periods of time during occupied hours is referred to as "duty cycling," and can be accomplished very easily with an EMS. Duty cycling saves fan and pump energy but does not reduce the energy required for space heating or cooling since the thermal demand must still be met.

The more sophisticated EMSs monitor the temperature of the conditioned area and use this information to automatically modify the duty cycle length when temperatures begin to drift. If, for example, the desired temperature in an area is 70° and at this temperature equipment is cycled 50 minutes on and 10 minutes off, a possible temperature-compensated EMS may respond as shown in Figure 11-1. As the space temperature increases above (or below if so programmed) the setpoint, the equipment off time is reduced until, at 80" in this example, the equipment operates continuously.

Duty cycling is best applied in large, open-space offices that are served by a number of fans. Each fan could be programmed so that the off times do not coincide, thereby assuring adequate air flow to the offices at all times.

Duty cycling of fans which provide the only air flow to an area should be approached carefully to ensure that ventilation requirements are maintained and that varying equipment noise does not annoy the occupants. Additionally, duty cycling of equipment imposes extra stress on motors and associated equipment. Care should be taken, particularly with motors over 20 hp, to prevent starting and stopping of equipment in excess of what is recommended by the manufacturer.

Peak Demand Charges

Electrical utilities charge commercial customers based not only on the amount of energy used (kWh) but also on the peak demand (kW) for each month. Peak demand is very important to the utility so they may

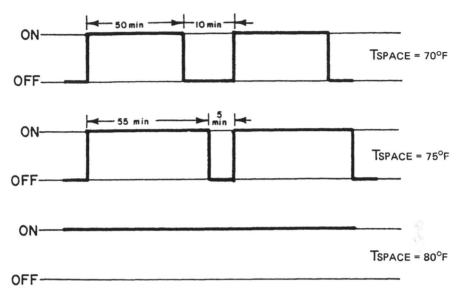

Figure 11-1. Temperature Compensated Duty Cycling

properly size the required electrical service and ensure that sufficient peak generating capacity is available to that given facility.

To determine the peak demand during the billing period, the utility establishes short periods of time called the "demand interval" (typically 15,30, or 60 minutes). The billing demand is defined as the highest average demand recorded during anyone demand interval within the billing period. (See Figure 11-2.) Many utilities utilize "ratchet" rate charges. A ratchet rate means that the billed demand for the month is based on the highest demand in the previous 12 months, or an average of the current month's peak demand and the previous highest demand in the past year.

Depending on the facility, the demand charge can be a significant portion, as much as 20%, of the utility bill. The user will get the most electrical energy per dollar if the load is kept constant, thereby minimizing the demand charge. The objective of demand control is to even out the peaks and valleys of consumption by deferring or rescheduling the use of energy during peak demand periods.

A measure of the electrical efficiency of a facility can be found by calculating the load factor. The load factor is defined as the ratio of energy usage (kWh) per month to the peak demand (kW) × the facility operating hours.

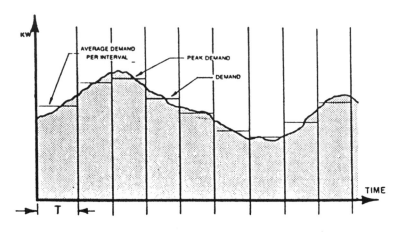

T - DEMAND INTERVAL

Figure 11-2. Peak Demand

Example SIM 11-2

What is the load factor of a continuously operating facility that consumed 800,000 kWh of energy during a 30-day billing period and established a peak demand of 2000 kW?

Answer

$$\text{Load Factor} = \frac{800,000 \text{ kWh}}{2000 \text{ kW} \times 30 \text{ days} \times 24 \text{ hours/day}} = 0.55$$

The ideal load factor is 1.0, at which demand is constant; therefore, the difference between the calculated load factor and 1.0 gives an indication of the potential for reducing peak demand (and demand charges) at a facility.

Demand Limiting

Energy management systems with demand limiting capabilities utilize either pulses from the utility meter or current transformers to predict the facility demand during any demand interval. If the facility demand is predicted to exceed the user-entered setpoint, equipment is "shed" to control demand. Figure 11-3 illustrates a typical demand chart before and after the actions of a demand limiter.

Electrical load in a facility consists of two major categories: essential loads, which include most lighting, elevators, escalators, and most production machinery; and non-essential ("sheddable") loads such as electric heaters, air conditioners, exhaust fans, pumps, snow melters, compressors and water heaters. When turned off for short periods of time to control demand, sheddable loads will not affect productivity or comfort.

To prevent excessive cycling of equipment, most energy management systems have a deadband that demand must drop below before equipment operation is restored (See Figure 11-4). Additionally, minimum on and maximum off times and shed priorities can be entered for each load to protect equipment and ensure that comfort is maintained.

It should be noted that demand shedding of HVAC equipment in commercial office buildings should be applied with caution. Since times of peak demand often occur during times of peak air conditioning loads, excessive demand limiting can result in occupant discomfort.

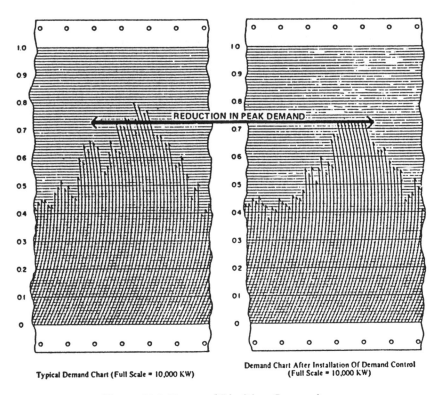

Typical Demand Chart (Full Scale = 10,000 KW)

Demand Chart After Installation Of Demand Control
(Full Scale = 10,000 KW)

Figure 11-3. Demand Limiting Comparison

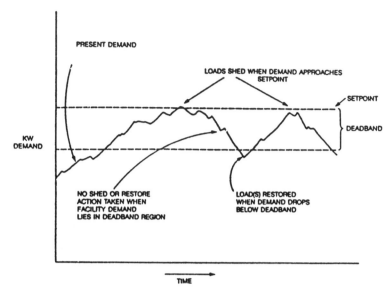

Figure 11-4. Demand Limiting Actions

Time-of-day Billing

Many utilities are beginning to charge their larger commercial users based on the time of day that consumption occurs. Energy and demand during peak usage periods (i.e., summer weekday afternoons and winter weekday evenings) are billed at much higher rates than consumption during other times. This is necessary because utilities must augment the power production of their large power plants during periods of peak demand with small generators that are expensive to operate. Some of the more sophisticated energy management systems can now account for these peak billing periods with different demand setpoints based on the time of day and day of week.

Optimal Start

External building temperatures have a major influence on the amount of time it takes to bring the building temperature up to occupied levels in the morning. Buildings with timer control usually start HVAC equipment operation at an early enough time in the morning (as much as three hours before occupancy time) to bring the building to temperature. During other times of the year, when temperatures are not as extreme, building temperatures can be up to occupied levels several hours before HVAC is necessary, and consequently unnecessary energy is used. (See

Figure 11-5.)

Energy management systems with optimal start capabilities, however, utilize indoor and outdoor temperature information, along with learned building characteristics, to vary start time of HVAC equipment so that building temperatures reach desired values just as occupancy occurs. Consequently, if a building is scheduled to be occupied at 8 :00 a.m., on the coldest day of the year, the HVAC equipment may start at 5 :00 a.m. On milder days, however, equipment may not be started until 7 :00 a.m. or even later, thereby saving significant amounts of energy.

Most energy management systems have a "self-tuning" capability to allow them to learn the building characteristics. If the building is heated too quickly or too slowly on one day, the start time is adjusted the next day to compensate. Artificial intelligence (AI) methods are also incorporated as part of an EMS.

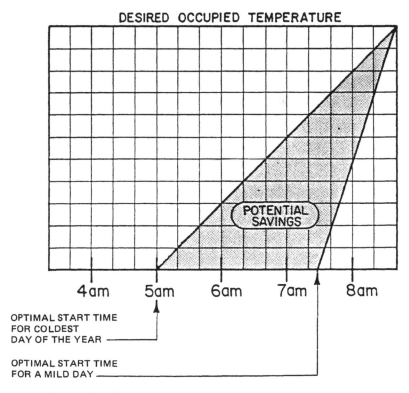

Figure 11-5. Typical Variation in Building Warm-up Times

Monitoring

Microprocessor-based EMS can usually accomplish a limited amount of monitoring of building conditions including the following:

- Outside air temperature
- Indoor air temperature (with several sensors)
- Facility electrical energy consumption and demand
- Several status input points

The EMS can store this information to provide a history of the facility. Careful study of these trends can reveal information about facility operation that can lead to energy conservation strategies that might not otherwise be apparent.

Direct Digital Control

The most sophisticated of the microprocessor-based EMSs provide a function referred to as "direct digital control" (DDC). This capability allows the EMS to provide not only sophisticated energy management but also basic temperature control of the building's HVAC systems.

Direct digital control is used in the majority of all process control applications and is an important part of the HVAC industry. Traditionally, pneumatic controls have been used in most commercial facilities for environmental control.

Control functions previously were performed by a pneumatic controller that received its input from pneumatic sensors (i.e., temperature, humidity) and sent control signals to pneumatic actuators (valves, dampers, etc.). Pneumatic controllers typically performed a single, fixed function that was not altered unless the controller itself was changed or other hardware was added. (See Figure 11-6 for a typical pneumatic control configuration.)

With direct digital control, the microprocessor functions as the primary controller. Electronic sensors are used to measure variables such as temperature, humidity and pressure. This information is used, along with the appropriate application program, by the microprocessor to determine the correct control signal, which is then sent directly to the controlled device (valve or damper actuator). (See Figure 11-6 for a typical DDC configuration.)

Direct digital control (DDC) has the following advantages over pneumatic controls:

- Reduces overshoot and offset errors, thereby saving energy.

- Flexibility to easily and inexpensively accomplish changes of control strategies.
- Calibration is maintained more accurately, thereby saving energy and providing better performance.

To program the DDC functions, a user programming language is utilized. This programming language uses simple commands in English to

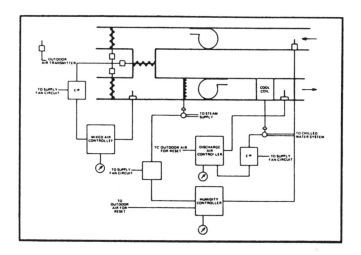

PNEUMATIC CONTROL SYSTEM

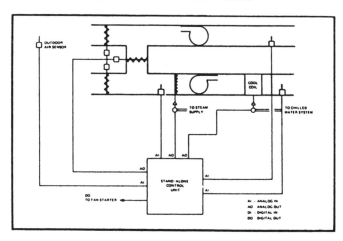

DIRECT DIGITAL CONTROL SYSTEM

Figure 11-6. Comparison of Pneumatic and DDC Controls

establish parameters and control strategies.

Computer-based EMS

A PC is used for operator interaction with the EMS. It accepts operator commands, displays data, and graphically displays systems controlled or monitored by the EMS. A printer, (or printers) provides a hard copy of system operations and historical data.

A "field interface device" (FID) provides an interface to the points that are monitored and controlled, performs engineering conversions to or from a digital format, performs calculations and logical operations, and accepts and processes commands. It is capable, in some versions, of stand-alone operations in the event of communications link failure.

The FID is essentially a microprocessor based EMS, as described previously. It may or may not have a keyboard/display unit on the front panel.

The FIDs are generally located in the vicinity of the points to be monitored/controlled. They are linked together (and to the PC) by a single twisted pair of wires which carries multiplexed data (i.e., data from a number of sources combined on a single channel) from the FID to the CPU and back. In some versions, the FIDs can communicate directly with each other.

Early versions of EMS used the central unit to perform all of the processing, with the FID used merely for input and output. A major disadvantage of this type of "centralized" system is that the loss of the central control disables the entire control system. The development of "intelligent" FIDs in a configuration known as "distributed control" helped to solve this problem. This system, which is prevalent today, utilizes microprocessor-based FIDs to function as remote controllers. Each panel has its own battery pack to ensure continued operation should the main controller fail.

Each intelligent unit sends signals back to the main unit only upon a change of status, rather than continuously transmitting the same value as previous "centralized" systems have done. This streamlining of data flow to the main unit frees it to perform other functions, such as trend reporting. The central unit's primary function becomes one of directing communications between various FID panels, generating reports and graphics, and providing operator interface for programming and monitoring.

Features

The primary advantage of the computer-based EMS is its increased capability to monitor building operations. For this reason, these systems are sometimes referred to as energy monitoring and control systems (EMCS). Analog inputs such as temperature and humidity can be monitored, as well as digital inputs such as pump or valve status.

The computer based EMS is also designed to make operator interaction very easy. Its operation can be described as "user friendly" in that the operator, working through the keyboard, enters information in English in a question and response format. In addition, custom programming languages are available so that powerful programs can be created specifically for the building through the use of simplified English commands.

The graphics display CRT can be used to create HVAC schematics, building layouts, bar charts, etc. to better understand building systems operation. These graphics can be dynamic so that values and statuses are continuously updated.

Many computer-based EMSs can also easily incorporate fire and security monitoring functions. Such a configuration is sometimes referred to as a building automation system (BAS). By combining these functions with energy management, savings in initial equipment costs can be achieved. Reduced operating costs can be achieved as well, by having a single operator for these systems.

The color graphics display can be particularly effective in pinpointing alarms as they occur within a building and guiding quick and appropriate response to that location. In addition, management of fan systems to control smoke in a building during a fire is facilitated with a system that combines energy management and fire monitoring functions.

Note, however, that the incorporation of fire, security and energy management functions into a single system increases the complexity of that system. This can result in longer start-up time for the initial installation and more complicated troubleshooting if problems occur. Since the function of fire monitoring is critical to building operation, these disadvantages must be weighed against the previously mentioned advantages to determine if a combined BAS is desired.

DATA TRANSMISSION METHODS

A number of different transmission systems can be used in an EMS for communications between the CPU and FID panels. These transmission systems include telephone lines, coaxial cables, electrical power lines, radio frequency, fiber optics and microwave. Table 11-1 compares the various transmission methods.

Table 11-1. Transmission Method Comparisons

Method	First Cost	Scan Rates	Reliability	Maint. Effort	Expandability	Compatibility with Future Requirements
Coaxial	high	fast	excellent	min.	unlimited	unlimited
Twisted pair	low	med.	very good	min.	unlimited	unlimited
RF	med.	fast but limited	low	high	very limited	very limited
Micro-wave	very high	very fast	excellent	high	unlimited	unlimited
Tele-phone	very low	slow	low to high	min.	limited	limited
Fiber optics	high	very fast	excellent	min.	unlimited	unlimited
Power Line Carrier	med.	med.	med.	high	limited	limited

Twisted Pair

One of the most common data transmission methods for an EMS is a twisted pair of wires. A twisted pair consists of two insulated conductors twisted together to minimize interference from unwanted signals.

Twisted pairs are permanently hardwired lines between the equipment sending and receiving data that can carry information over a wide range of speeds, depending on line characteristics. To maintain a particular data communication rate, the line bandwidth, time delay, or the signal-to-noise ratio may require adjustment by conditioning the line.

Data transmission in twisted pairs, in most cases, is limited to 1200 bps (bits per second) or less. By using signal conditioning, operating speeds up to 9600 bps may be obtained.

Voice Grade Telephone Lines

Voice grade lines used for data transmission are twisted pair circuits. Two of the major problems involve the quality of telephone pairs provided to the installer and the transmission rate.

The most common voice grade line used for data communication is the unconditioned type. This type of line may be used for increased data transmission, with the proper line conditioning. Voice grade lines must be used with the same constraints and guidelines as twisted pairs.

Coaxial Cable

Coaxial cable consists of a center conductor surrounded by a shield. The center conductor is separated from the shield by a dielectric. The shield protects against electromagnetic interference. Coaxial cables can operate at data transmission rates in the megabits per second range, but attenuation becomes greater as the data transmission rate increases.

The transmission rate is limited by the data transmission equipment and not by the cable. Regenerative repeaters are required at specific intervals, depending on the data rate (nominally every 2000 feet), to maintain the signal at usable levels.

Power Line Carrier

Data can be transmitted to remote locations over electric power lines using carrier current transmission that superimposes a low power RF (radio frequency) signal, typically 100 kHz, onto the 60 Hz power distribution system. Since the RF carrier signal cannot operate across transformers, all communicating devices must be connected to the same power circuit (same transformer secondary and phase) unless RF couplers are installed across transformers, permitting the transmitters and receivers to be connected over a wider area of the power system. Transmission can be either one-way or two-way.

Note that power line carrier technology is sometimes used in microprocessor based EMS retrofit applications to control single loads in a facility where hardwiring would be difficult and expensive (e.g., wiring between two buildings). Figure 11-7 shows a basic power line carrier system configuration.

Radio Frequency

Modulated RF signals, usually VHF or FM radio, can be used as a data transmission method with the installation of radio receivers and transmitters. RF systems can be effectively used for two-way communication between CPU and FID panels where other data transmission methods are not available or suitable for the application. One-way RF systems can be effectively used to control loads at remote locations such as warehous-

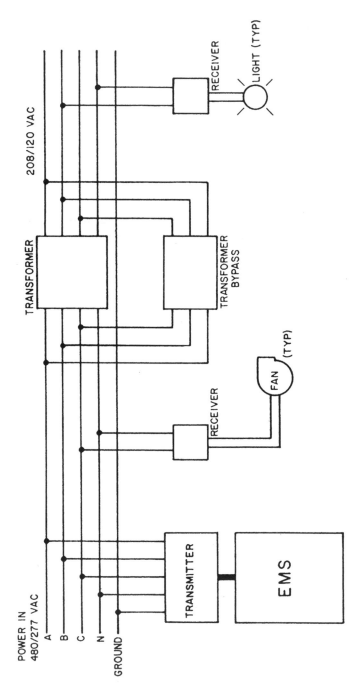

Figure 11-7. Power Line Carrier System Configuration

es, unitary heaters, and family housing projects.

The use of RF at a facility, however, must be considered carefully to avoid conflict with other existing or planned facility RF systems. Additionally, there may be a difficulty in finding a frequency on which to transmit, since there are a limited number available.

The kinds of signals sent over an FM radio system are also limited, as are the distances over which the signals can be transmitted. The greater the distance, the greater the likelihood that erroneous signals will be received.

Fiber Optics

Fiber optics uses the wideband properties of infrared light traveling through transparent fibers. Fiber optics is a reliable communications media that is rapidly becoming cost competitive when compared to other high-speed data transmission methods.

The bandwidth of this media is virtually unlimited, and extremely high transmission rates can be obtained. The signal attenuation of high-quality fiber optic cable is lower than the best coaxial cables. Repeaters required nominally every 2000 feet, for coaxial cable, are 3 to 6 miles apart in fiber optics systems. Fiber optics must be carefully installed and cannot be bent at right angles.

Microwave Transmission

For long distance transmission, a microwave link can be used. The primary drawback of microwave links is initial cost. Receivers/transmitters are needed at each building in a multi-facility arrangement.

Microwave transmission rates are very fast. Reliability is excellent, too, but knowledgeable maintenance personnel are required. The only limit on expansion is cost.

SUMMARY

The term energy management system denotes equipment whose functions can range from simple timer control to sophisticated building automation.

Capabilities of EMS can include scheduling, duty cycling, demand limiting, optimal starting, monitoring, direct digital control, fire detection, and security. Direct digital control capability enables the EMS to replace the environmental control system so that it directly manages HVAC operations.

An energy management system (EMS) may be used for energy management in control of building systems, groups of buildings, and in power transmission and distribution.

A building systems EMS often refers to a computer system designed specifically for the automated control and monitoring of the heating, ventilation and lighting needs of a building or group of buildings. Most of these energy management systems provide reading of utility meters for electricity, gas and water. Trend analysis and consumption forecasts can be made from the data obtained.

A power transmission EMS is usually a computer based system used by operators of electric *utility grids* to monitor, control, and optimize the performance of the *generation* and/or *transmission system*. SCADA (an acronym for supervisory control and data acquisition) is used to monitor and control functions. Used with SCADA systems are optimization packages for advanced applications. These systems are referred to as EMS/SCADA.

Chapter 12

Power Line Disturbances: A User's Perspective on the Selection and Application of Mitigation Equipment Techniques*

INTRODUCTION

The proliferation of personal computers (PCs), programmable logic controllers (PLCs), energy management systems (EMS), microprocessor-based instrumentation, and other state-of-the-art electronic devices in an industrial plant setting has prompted a new awareness of the impact of power line disturbances (PLD) on the reliability of these systems and the ability to perform the desired tasks in an interruption-free manner. While it is common to consider the local power company as the source of all PLD problems, investigation has revealed that in many cases a significant portion originate in the plant itself, from adjustable speed drives, heater and lighting controls, copy machines, and even the switching power supplies of computers and controllers themselves. This chapter reviews the various types of power line disturbances and their possible sources.

All dollar values presented are approximate and are based on energy and equipment costs in the Southeast. Brand names are included to present a cross section of equipment available, and the listing is not intended to be all inclusive.

---

*Presented at 12th World Energy Engineering Congress by W.L. Stebbins, sponsored by The Association of Energy Engineers, Atlanta, GA.

There are a variety of firms capable of supplying satisfactory equipment and services. The suppliers used were chosen based on their previous experience with similar applications.

Application of this equipment and services for these specific requirements should not be construed as a general endorsement by either Hoechst Celanese Corporation or by the author. It is important to note that a variety of brands and suppliers should be evaluated by anyone considering a similar application.

Background

The challenge is to provide the required power quality for these systems while meeting corporate economic guidelines.

POWER QUALITY DISTURBANCES

Specifications

Specifications for commercial and industrial power are adequate for operating motors, heaters, and lighting, which constitute more than 95% of the load base. Electronic loads require cleaner power. Power quality disturbances such as voltage sags, surges, and impulses; faults in the shape of the voltage sine wave; high frequency noise; harmonics; and frequency errors can all cause intermittent failures of electronic loads. These same power disturbances can also shorten the expected life of an electronic device by causing increased heating and higher electrical stress.

Although it is technically possible to construct electronic loads that tolerate virtually any kind of power quality disturbance, the cost of construction makes this approach impractical. For example, in a typical electronic instrument the power supply is 5% of the cost; but in a power disturbance monitoring instrument designed to tolerate almost any power disturbance, the power supply's cost rises to as much as 30%.

It is also possible to construct power conditioning systems that clean up power before it is presented to the electronic load and eliminate virtually all power disturbances. This approach is costly and is effective only when the load is concentrated in one physical location (such as a room dedicated to a mainframe computer). It is more difficult and even more costly when the loads are distributed throughout the building or entire industrial plant facility.

Table 12-1 illustrates typical electronic requirements vs. typical utility delivery specifications.[1]

Table 12-1. Electronic Requirements vs. Utility Specifications

Power Disturbance Event	Typical Electronic Load Requirements	Typical Utility Delivery Specification
Voltage surge	No more than 10% for no longer than 32 milliseconds	No more than 20% for no longer than 60 ms
Voltage sag	No more than 13% for no longer than 200 milliseconds	No more than 20% for no longer than 500 ms
Impulse	No more than 200 Vpk for no longer than 20 microseconds	Not specified
Noise	No more than 5 V pk-to-pk between 100 KHz and 2 MHz	Not specified

(The utility delivery specification applies at the service entrance to the building. Typically, the customer's loads cause larger disturbances where the electronic load is located.)

Responsibility for Power Quality

The end-user is responsible for providing an acceptable operating environment for the electronic load. His perception of his responsibility is distorted by two factors: sales implications that the equipment will operate from "ordinary" power, and the premise that it is the equipment field service organization's responsibility to keep his system running.

Many electronics suppliers have a weak or non-existent pre-site inspection requirement, even to the point of stating their electronic loads will operate from "standard" power. This may be phrased as "office power," "a standard outlet," or even "a dedicated circuit." Other suppliers explicitly state their power requirements, including tolerance levels for common power quality faults, However, explicit power quality requirements are virtually impossible to measure unless the end-user or field service engineer is equipped with the proper training, procedures, and instrumentation.

POWER LINE DISTURBANCES—CAUSE AND EFFECT

One of the biggest obstacles facing anyone attempting to understand power line transient phenomena is the language itself, A great deal of jargon is used in the engineering profession and in the instrumentation and computer industry to describe power line disturbances. Also, if any action plan is to be formulated to deal with these disturbances, it is important to know something about their sources and the effects they have on equipment and computer hardware.[2]

Voltage Variation

A voltage variation refers to any change or swing in steady-state voltage above or below the prescribed input range for a given piece of equipment. Voltage variations include brownouts, which are intentional reductions by the utilities to conserve energy for a specific period of time. Voltage variations result in improper function of logic and memory circuits, and overheating in the case of voltage increases. Voltage variations are caused by unregulated utility feeders that experience changes in load over a period of time. Sometimes they may occur when utilities switch their systems to change the way a particular customer is served, Voltage variations may also frequently occur within a building or industrial plant due to load changes occurring within the premises, even when the utility voltage is constant, and even in cases when on-site power generation exists.

Powerfail

Powerfail is defined as the total removal of input voltage for at least 5 milliseconds. Powerfails can cause the floating heads of disc drives to crash down on the disc, causing memory loss, unscheduled shutdown, or equipment damage. Some disc drives have heads which automatically retract upon loss of power, but it is not safe to assume this without checking the equipment specifications. Powerfails can also cause improper operation of logic and memory circuits. The length of powerfail a computer or other equipment can tolerate will depend on the size of the backup logic battery onboard, or the ride-through provided by the LC filtering circuit of the power supply, and the load on the power supply at the time of the powerfail. Some power supplies provide up to 50 milliseconds (3 cycles at 60 HZ) of ride-through. Often large AC motors connected to the same power system will act as generators for a few cycles to provide additional ride-through.

Powerfails result from utility switching operations and equipment failures, Proper operation of lightning arrestors cause short (one cycle or less) powerfails. The most common powerfail is that caused by utility reclosing circuit breakers acting to clear lightning-induced flashover, or a fault caused by a powerline coming in contact with a tree branch or other grounded object. Most breakers take a minimum of 30 to 45 cycles to open and reclose, which is far beyond the ride-through capability of most computer power supplies. Large SCRs used in DC and adjustable-speed AC motor drives can "notch" the sine wave, causing what amounts to a short powerfail.

Transients

The term transient is used to define a high amplitude, short duration disturbance of from less than one microsecond to several milliseconds that is superimposed on the normal voltage wave. This term is also often used loosely to describe any disturbance that is transitory, such as common-mode noise, surges, sags, and other phenomena. Transients are caused by lightning, capacitor switching, fault switching, arcing grounds, brush-type motors such as drills and office machines, and the switching of inductive loads such as motors, transformers, lighting ballasts, x-ray equipment, and solenoids.

The effect of transients on the computer can be data errors, due to the spike voltage passing through the interwinding capacitances of the power supply and on to the logic circuits. Actual damage to the equipment can result if a very high voltage transient occurs.

Surges and Sags

Of all the jargon used to describe power system disturbances, the word "surge" is probably the most misunderstood. In the utility industry, the word surge is used to describe transients of less than 1/2 cycle in duration. The IEEE defines a surge as a transient wave of voltage or current. Computer manufacturers use the term surge to describe a sudden increase in voltage of more than 1/2 cycle but less than a few cycles. Of all the types of power line disturbances, surges are perhaps the least common and the least troublesome. Surges result from power system malfunctions and sudden load changes, such as removing a large motor from the line.

A sag is a sudden reduction in voltage greater than 1/2 cycle in duration, and it is most frequently caused by faults being cleared on the utility line and the starting of large motors across-the-line within the plant. Sags mayor may not be troublesome to computers and other electronic devices, depending on the regulating ability of the power supply and its ride-through ability.

Harmonic Distortion

One of the greatest causes of harmonic distortion in today's power systems is SCR switching. This distortion can be created by large solid-state power supplies for AC and DC drive motors. Often the computer power supply is the worst offender. Whatever the specific cause, the equipment being protected and the electrical environment in the vicinity of the protected equipment will determine the harmonic content.

Motor, incandescent lighting, and heating loads are linear in nature. That is, the load impedance is essentially constant regardless of the applied voltage. For alternating current, the current increases proportionately as the voltage increases, and it decreases proportionately as the voltage decreases. This current is in phase with the voltage for a resistive circuit with a power factor (PF) of unity.

In the past, most loads were linear, and the non-linear loads were a small portion of the total as to have little effect on the system design and operation. Then came electronic loads such as computers, UPS equipment, and adjustable-speed motor drives. These electronic loads are mostly nonlinear and are a large factor to have serious consequences on distribution systems. Overheated neutral conductors, failed transformers, malfunctioning generators, and motor burnouts have been experienced, even though loads were apparently well within equipment ratings.

A nonlinear load is one in which the load current is not proportional to the instantaneous voltage. Often, the load current is not continuous. It can be switched on for only part of the cycle, as in a thyristor-controlled circuit, or pulsed, as in a controlled rectifier circuit, a computer, or power to a UPS. The major effect of nonlinear loads is to create considerable harmonic distortion on the system, These harmonic currents cause excessive heating in magnetic steel cores of transformers and motors. Odd-order harmonics are additive in the neutral conductions of the system, and some of the pulsed currents do not cancel out in the neutral, even when the three phases of the system are carefully balanced. The result is overloaded neutral conductors. Nonlinear load currents are non-sinusoidal, and even when the source voltage is a clean sine wave, the nonlinear loads will distort that voltage wave, making it non-sinusoidal.[3,4]

In addition, harmonic distortion typically created by *non-linear loads* can cause problems when power factor correcting capacitors are added. The factor harmonics cause resonance and high values often appear seemingly at random when certain combinations of equipment are run.

A "tuning" inductor is added in series with the p.f. correcting "C" to "tune out" the distorting harmonics. A "tuning" inductor added in series with the p.f. correcting capacitor will "tune out" the harmonics.

CATEGORIES OF POWER LINE DISTURBANCES

It is useful to define power line disturbances into Types I, II, and III depending on their duration.

Type I
Transients, sags, short powerfails, and other disturbances of less than 1/2 cycle are classified as Type I. Many studies have been done showing that Type I disturbances are very common on the low-voltage power system. The effect of these disturbances on computers is subject to debate; only the most severe transient overvoltages are likely to disturb modern-day electronic devices.

Very short duration overvoltages that have little total energy will usually be absorbed by line filter elements and surge arresters (if any) before reaching the components most susceptible to breakdown. Experience indicates that there are few problems with most computers resulting from AC power source overvoltages. It is most difficult for the extra energy to pass through filters and regulators and cause the regulated power supply voltages to increase significantly at the logic circuit chips. The voltage transient associated with the overvoltages may find other paths, however, to interact with data signals and cause data corruption.

Type II
Type II disturbances include surges, sags, and powerfails of 1/2 cycle to 2 seconds. Momentary undervoltage (sags) resulting from power system overloads or fault clearing are the most common causes of power-related electronic equipment system failures. Weather is one of the major causes of momentary undervoltages and powerfails in this category.

Type III
Type III disturbances include voltage variations, brownouts, and outages longer than two seconds. Planning for this category of disturbance includes ride-through logic batteries, full UPS systems, and diesel-powered generator sets.

MONITORING AND ANALYZING EQUIPMENT

Equipment for diagnosing power line disturbances can be purchased at prices ranging from only a few hundred to thousands of dollars, and each type has its place. Many facilities may find it feasible to have a variety of simple, low-cost units to identify problems concurrently at various locations in the plant, without tying up several more expensive instruments. One or more comprehensive instruments might be kept in the maintenance instrument inventory to provide qualitative information on the specific nature of the problem after the lower-cost units have determined that a disturbance exists.

Opinions differ among manufacturers' claims that analysis of disturbances is an expensive waste of time, citing that 95 percent of electronic equipment misoperations and failures are caused by voltage transients of some duration that can be practically eliminated by simply installing a low-cost transient suppressor. On the other hand, manufacturers of UPS systems might equally claim that installing a UPS system will eliminate 100 percent of power line disturbances.

MITIGATION EQUIPMENT AND TECHNIQUES

Dedicated Circuit

The request for a dedicated circuit has always been a source of challenge in engineering offices. The ideal dedicated circuit will have, within reason, the lowest possible impedance. This is true because computers draw higher instantaneous currents during turn-on. To achieve this, the source of the circuit should theoretically be as close to the building service entrance as possible. However, it is not safe to assume that this will provide disturbance-free power, and in many cases the reverse is true, because transient spikes are traveling waves and the electronic devices should be as far away from the source of transients as possible. Thus, the point of connection of the "dedicated circuit" is a judgment call. Also, in a typical industrial plant of several million square feet, it is just not practical to run dedicated circuits to each of several hundred computers, terminals, and other electronic devices.

Shielded Isolation Transformers

Next to the dedicated circuit, the shielded isolation transformer is the most popular power conditioner. Grounding the secondary of the delta-wye isolation transformer provides the computer with a clean, noise-free ground. Most power conditioner manufacturers agree this is among the most important factors in providing a trouble-free power environment. The shielded isolation transformer also provides excellent common-mode noise rejection, because of the low capacitance between the primary and secondary windings. Capacitive or electrostatic coupling is the only way common-mode noise can be transmitted from primary to secondary. There is no magnetic coupling, because common-mode voltages do not impress any line-to-line or line-to-neutral voltages across the primary windings. (See Figure 12-1.)

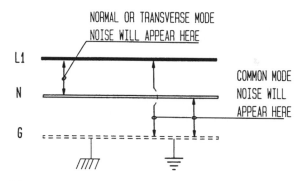

Figure 12-1. Types and Locations of Electrical Noise

The shield, which usually consists of a foil wrapping of conducting non-magnetic material, conducts the electrostatic charge around the primary winding to ground, preventing it from coupling with the secondary, while providing greater than 300-to-1 common-mode noise reduction.

Isolation transformer shields are generally ineffective in rejecting transverse (normal) mode transients, although there is some attenuation. Transverse mode transients or noise (unlike common-mode) do impress a voltage across the primary windings. The isolation transformer is also useless in protecting against surges, sags, dips, and variations in steady-state voltage. The shielded isolation transformer is often used in tandem with voltage regulators, transient suppressors, and other devices and power conditioners, because it offers excellent common-mode noise rejection and a clean ground.

Voltage Regulators

Constant voltage transformers (CVT) using ferroresonant technology are one of the most popular types of voltage regulators. These devices utilize a ferroresonant circuit consisting of a capacitor in series with a transformer coil. The saturated core provides immunity to input voltage variations. Because of the saturated core, secondary voltage remains fairly constant, in spite of changes in primary voltage. As a result, this device takes virtually no time to respond. One common use is to power the primary logic motherboard in an adjustable speed drive from a CVT, to help isolate it from voltage variations.

However, these devices are known to have high impedances and to be sensitive to changes in load. Therefore, high inrush loads such as switching power supplies can produce transients and noise on the output. When electrostatic shielding and special transformer geometries are used, ferroresonant devices can provide excellent common-mode noise rejection.

Transient Suppressors

Transverse-mode transient spikes are voltage spikes between line conductors. These are produced by the power system through switching, SCR commutation action, faults, and grounds. As mentioned earlier, the spike is generally the first few half cycles of the leading edge of an oscillating overvoltage. Probably the least expensive, although not the most effective, way to remove these overvoltages is through surgery, simply cutting them off or clipping them. This is the least understood and most misapplied technique.

Probably the most common low voltage (120 volts), transient suppressor is the metal oxide varistor (MOV), which is usually fast enough to clamp most transients. Silicon avalanche diodes are used to achieve very fast (5 nanosecond) clamping. The use of surge suppressors, capacitors, and LC filters to eliminate spikes and their accompanying noise is fairly simple and straightforward.

Keep in mind that, in general, these devices only clip the spikes to ± several hundred volts peak-to-peak, which can still cause problems for PCs and other electronic equipment.

One of the disadvantages of transient suppressors is that they pump transients to ground, converting a line problem to a ground potential problem. This can effectively convert a transverse mode problem to a common-mode problem. Surge suppressors and lightning arrestors should always be separated from the computer or electronic device ground by an isolation transformer.

One very unusual and highly accurate "active tracking" filter senses the instantaneous sine wave voltage at any point in the cycle. The LC circuit limits the maximum voltage that can appear at the output of the filter. This filter begins to limit the deviation from the true instantaneous sine-wave voltage whenever the deviation becomes greater than ± 2 volts. When a voltage deviation is sensed, the unit switches the capacitor leg into the circuit, providing full filtering in less than 5 ns. Not only are spikes clipped very effectively, but the trailing notches are also "filled in" by the energy stored in the filter's capacitance, which cannot be done by simple clamping devices.[5] (See Figure 12-2.)

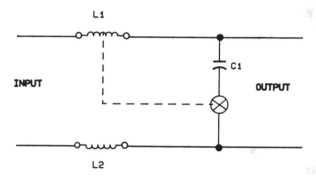

Figure 12-2. Active Tracking Filter

These devices are available in receptacle models and hard-wired models, both single- and three-phase, in 120, 240, and 480 volt styles.

An excellent discussion of transient suppressors and methods of testing their effectiveness is detailed in reference [6]. The reader is also directed to UL 1449 Standard, Transient Surge Suppression.

Uninterruptible Power Supply

Where a high degree of voltage regulation and total isolation from sags, dips, surges, and transients is required, an uninterruptible power supply (UPS) may be used. UPS provides standby capacity for powerfails and short outages and provides constant frequency. As a minimum, the UPS allows for orderly shutdown of computer equipment without risk of equipment damage or memory loss. Since UPSs are solid-state devices, they often require power conditioners themselves. A surge suppressor or other type of transient remover is necessary to prevent damage to their solid-state components. Depending on the kVA rating, a UPS unit can take up a great deal of space, and because of the weight of the batteries, care must be taken in selecting a battery room that can handle the structural load.

Perhaps the biggest disadvantage of using UPS is their high impedance and the high voltage drops associated with that impedance. This is of particular importance with switching power supplies, which themselves are noise generators. Sometimes it is necessary to size these systems for one or more multiples of normal operating current, to compensate for the resulting voltage drop. Since UPS are more expensive than other forms of power conditioners, especially when they are required to be oversized, the user should be sure they are really needed before making the investment.

There are three major UPS versions. (1) Continuous service batteries. Direct current from the batteries is inverted to AC to power the UPS changes utility power to direct current to charge a set of page computer. (2) A standby type of UPS involves operating the computer on line power, but it switches to battery power in case of main power interruption. This can cause a very short (4 to 16 ms) power dip to the load. The battery is continuously kept charged by the battery charger. (3) A reverse transfer UPS offers additional security, because the computer load is transferred to utility power if the UPS is temporarily overloaded or when a malfunction occurs within the UPS itself.

It is important to realize that the output distortion of the UPS for a given load depends on the UPS design and output impedance. If the load distorts the power supplied by the UPS, then the power fed to the loads will not be truly "clean." Most UPS manufacturers specify the output distortion of their equipment; 5% total harmonic distortion (THD) is typical. However, many manufacturers add a disclaimer, such as "based on linear loads" or "for reactive and inductive loads." Such a disclaimer means that the THD figure only applies under linear load conditions. Before purchasing any UPS, make certain that it is capable of supplying the actual types of nonlinear loads to be connected.

Less expensive ride-through of up to half a second can be achieved when using adjustable speed drives by adding capacitors to the DC bus, providing the power supply is suitably sized. This approach was recently used effectively when applied to a system of HVAC fans in a large industrial building[7,8]. The IEEE Orange Book serves as the industry standard for information of UPS and standby power systems.[9]

Grounding

No discussion on PLD is complete without giving consideration to power system and instrumentation grounding requirements. Grounding must serve many purposes, not all of which are simultaneously compatible. The simplest task served by a system ground is the establishment of an electrical potential reference. The problems arise when many system references must be connected together.

The individual instruments and computer components usually have all exposed metal parts connected to the power supply ground. During faults on the utility lines, potential differences of several hundred volts may be generated between the equipment and the substation ground up to several thousand feet away. On a less spectacular scale, leakage currents may produce potential differences of many millivolts between the cases of instruments within the same room, causing errors in the proper operation of the devices.

Several well-known publications can provide guidance when reviewing the various considerations to be included in any grounding scheme.[10,11,12]

Conclusion

The intent of this chapter has been to establish that the selection and application of power conditioning equipment is a complex issue with no simple solution. Sometimes even the "experts" disagree in their basic philosophies. They disagree on the nature of the problem, and they disagree as to which technologies are best for solving the problem. Selection of power conditioning equipment requires an understanding of power system disturbances and the available technologies for eliminating them.

It is also important to understand the consequences and the cost of the downtime, unscheduled outages, and equipment damage. No conditioner should ever be applied to any piece of equipment without first consulting the manufacturer who knows for sure what types of disturbances are likely to damage or otherwise affect that particular brand of equipment.

Providing just the right level of protection to each PLD application is the opportunity, and as pointed out at the beginning of this chapter, that is your challenge.

References

1. Basic Measuring Instruments 402 Lincoln Center Drive Foster City, CA 94404-1161
2. Hugh O. Nash, Jr., and Frances M. Wells, "Power Systems Disturbances and Considerations For Power Conditioning," IEEE Industrial & Commercial Power System Technical Conference Paper.
3. Arthur Freund, "Nonlinear Loads Mean Trouble,: EC&M, Pages 83-90.
4. Arthur Freund, "Double The Neutral and Derate The Transformer—Or Else!" EC&M, Pages 81-85.
5. Arthur Freund, "Protecting Computers from Transients," EC&M, pages 65-70
6. Surge Protection Test Handbook KPS-109, Keytek Instrument Corp., 260 Fordham Road, Wilmington, Mass. 01887
7. W.L. Stebbins, "A User's Perspective On The Application Of Adjustable Speed Drives and Microprocessor Control for HVAC Savings," IEEE Textile Industry Technical Conference Paper.
8. Southern Industrial Controls, 10901 Downs Road, P.O. Box 410328, Charlotte, NC 28241-0328
9. IEEE Recommended Practice For Emergency And Standby Power For Industrial And Commercial Applications (IEEE Orange Book) ANSI/IEEE STd. 446-1987.
10. IEEE Recommended Practice For Grounding of Industrial and Commercial Power Systems (IEEE Green Book) ANSE/IEEE Std. 142-1982.
11. IEEE Recommended Practice For Electric Power Distribution For Industrial Plants, Chapter 7, (IEEE Red Book), ANSI/IEEE Std. 141-1986.
12. Guidelines On Electrical Power For ADP Installations, Chapter 3, Federal Information Processing Standards Publication, (PIPS PUB 94), U.S. Department of Commerce, National Bureau of Standards.

Chapter 13

Variable Speed Drives

INTRODUCTION

Variable speed drives can be used in conjunction with supply and return fans and cooling tower fans, as well as with virtually any type of centrifugal pump. In addition, centrifugal chillers are a prime candidate. Speed control is considered primarily for its energy savings benefits.

VARIABLE SPEED DRIVES

There are several variable speed drive systems that are available that have been used for control of fan or pump speed. These include hydroviscous, eddy current clutch, direct current, and variable frequency converters.

The unique advantage of the adjustable frequency drive is that the standard AC motors may be used. This means that equipment such as cooling towers can be converted to automatic control without any change in the mechanical components.

This chapter will highlight AC variable frequency drive applications.

INDUCTION MOTORS

An induction motor is a constant -speed device. Its speed depends on the number of poles provided in the stator, assuming that the voltage and frequency of the supply to the motor remain constant.

Several methods can be used to verify the speed of an AC motor. The stator or primary winding can be connected to change the number of poles. For example, reconnecting a 4-pole winding so that it becomes a 2-pole winding will double the speed. This method can give specific alternate speeds but not gradual speed changes. Or, the slip can be changed for a given load by varying the line voltage. However, torque is proportional to the square of the voltage, so reducing the line voltage rapidly reduces the available torque and will soon cause the motor to stall. Only very limited speed control is possible by this method.

An excellent way to vary the speed of a squirrel-cage induction motor is to vary the frequency of the applied voltage. To maintain a constant torque, the ratio of voltage to frequency must be kept constant, so the voltage must be varied simultaneously with the frequency. Adjustable frequency controls perform this function. At constant torque, the horsepower output increases directly as the speed increases.

For a 60-Hz motor, increasing the supply frequency above 60 Hz will cause the motor to be loaded in excess of its rating, which must not be done except for brief periods. For a supply frequency of less than 60 Hz, the speed will be less than the design speed of the motor. As the frequency is reduced, the voltage should also be reduced, to maintain a constant torque. Sometimes it is desirable to have a constant horsepower output over a given speed range. These and other modifications can be obtained by varying the ratio of voltage to frequency as required. Some controllers are designed to provide constant torque up to 60 Hz and constant hp above 60 Hz to permit higher speeds without overloading the motor.

The speed of an AC induction motor can be changed over a very wide range, from perhaps 10% to 20% of 60-Hz-rated speed up to several times rated speed. However, several cautions must be observed. At higher speeds, care must be taken not to exceed the hp rating of the motor. At speeds more than 10% above rating, the manufacturer must be consulted as to the ability of the motor to withstand the mechanical forces involved. At low speeds, roughly 20% of rated speed or less, especially if the motor is fan-cooled, care must be taken not to exceed the permitted motor temperature rise. If speed gets too low, the motor may "cog"—the rotor jumping from one position to the next instead of rotating smoothly—or it may stall completely.

Capability, versatility and flexibility of the AC induction motor are matters of fact. To obtain maximum suitability and effectiveness when selecting and applying AC induction motors, many other factors must be considered. These include type of application, enclosure, mountings, coupling, bearings, insulation, temperature ratings, initial costs, operating cost, energy rating, and starting and control requirements.

HOW SPEED CONTROLLERS WORK

The variable frequency drives (VFD), which act as an interface between the AC power supply and the induction motor, must achieve the following basic requirement:*

1. Ability to adjust the frequency according to the desired output speed.
2. Ability to adjust the output voltage so as to maintain a constant air-gap flux in the constant torque region.
3. Ability to supply a rated current on a continuous basis at any frequency.

Figure 13-1 illustrates the basic concept where the AC power input is converted into DC by means of either a controlled or an uncontrolled rectifier. The DC link filter the ripple at the output of the rectifier, and the combination of the controlled rectifier and filter provides a variable DC voltage to the inverter. The inverter converts DC to variable frequency AC. An inverter belongs to the voltage source. Similarly, an inverter which behaves as a current source at its terminal is called a current source inverter.

Because of a low internal impedance, the terminal voltage of a voltage source inverter remains substantially constant with variations in load. It is therefore equally suitable to single-motor and multimotor drives. Because of a large internal impedance, the terminal voltage of a current

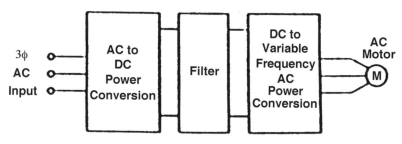

Figure 13-1. Variable Frequency Converter

---

*Applicability of Variable Frequency Control Is Not a General Statement," M. Karmous, G. Smith, Presented at the 12th World Energy Engineering Congress, Sponsored by the Association of Energy Engineers, Atlanta, GA.

source inverter changes substantially with a change in load.

An AC motor is designed to operate at a specific voltage and operating frequency determined at the time of manufacture, for example 460 volts/AC (VAC) at 60 hertz (Hz). This voltage-to-frequency relationship is referred to as volts per hertz so V/Hz is derived by dividing voltage by frequency.

$$460 \text{ VAC}/60 \text{ Hz} = 7.66 \text{ V}/\text{Hz} \qquad \text{(Formula 13-1)}$$

For proper operation and full available motor torque, this ratio must be maintained over the entire operating sped range of the system. Therefore, if we wish to reduce the speed by one half (to 30 Hz), we must reduce the voltage by one half (to 230 VAC).

$$230 \text{ VAC}/30 \text{ Hz} = 7.66 \text{ V}/\text{Hz} \qquad \text{(Formula 13-2)}$$
(The V/Hz ratio is maintained)

VOLTAGE SOURCE INVERTER

The power circuit of a three-phase voltage source inverter (VSI) is shown in Figure 13-2. It consists of six self-commutation switches S_1 to S_6 with the anti parallel diodes D_1 to D_6. The switches need not have reverse voltage blocking capability. They may be realized using power transistors, GTOs, MOSFETS, or inverter grade thyristors with forced commutation circuits. For diodes D_1 to D_6, fast recovery diodes are employed. A snubber is required for each switch-diode pair. The motor, which is connected across terminal A, B, and C, may have wye or delta connection. The inverter may be operated as a square-wave inverter or as a pulse-width modulated (PWM) inverter.

Square-wave VSI Drives

The inverter operates in a square-wave mode, which results in phase-to-motor neutral voltage as shown in Figure 13-2(b). With the square-wave inverter operation, each inverter switch is on for 180°, and a total of three switches are on at any instant of time. The resulting motor current waveform is also shown in Figure 13-2(b), Because of the inverter operating in a square-wave mode, the magnitude of the motor voltage is controlled by controlling V_a in Figure 13-2(a) by means of a line-frequency phase-controlled converter.

Voltage harmonics in the inverter output decrease as V_1/h with $h = 5, 6, 11, 13$, etc., where V_1 is the fundamental frequency phase-to-neutral voltage. Because of substantial magnitudes of low-order harmonics, these currents result in large torque ripple, which can produce troublesome speed ripple at low operating speeds.

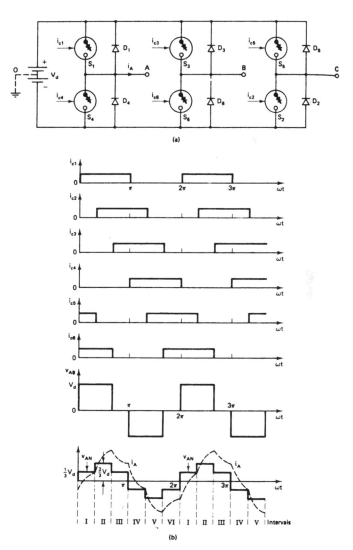

Figure 13-2. 3-φ Voltage Source Inverter

PWM-VSI Drives

Figure 13-3 shows the schematic of a pulse-width-modulated PWM-VSI drive. Assuming a three-phase utility input, a PWM inverter controls both the frequency and the magnitude of the voltage output. Therefore, at the input, an uncontrolled diode bridge rectifier is generally used. In a PWM inverter, the harmonics in the output voltage appear as sidebands of the switching frequency and its multiples. Therefore, a high switching frequency results in an essentially sinusoidal current in the motor.

Since the ripple current through the DC-bus capacitor is at the switching frequency, the "input DC source" impedance seen by the inverter would be smaller at higher switching frequencies. Therefore, a small value of capacitance suffices in PWM inverters, but this capacitor must be able to carry the ripple current. A small capacitance across the diode rectifier also results in a better input current waveform drawn from the utility source: However, care should be taken to avoid letting the voltage ripple in the dc-bus voltage become too large, which would cause additional harmonics in the voltage applied to the motor.

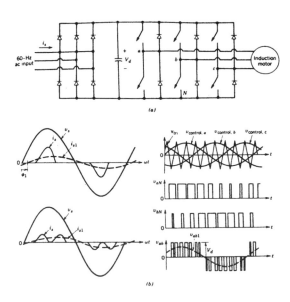

Figure 13-3. PWM-VSI a) schematic, b) waveforms

VARIABLE FREQUENCY DRIVE APPLICATIONS*

VFDs are the most efficient means of capacity control for centrifugal fans and pumps. They control a motor's speed electronically rather than mechanically. Figure 13-4 shows that a VFD uses nearly the theoretical minimum power predicted by the centrifugal fan and pump affinity laws. In effect, the VFD sizes the motor to the load. The speed of an AC motor

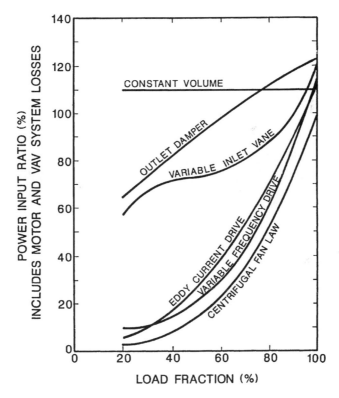

Figure 13-4. Typical Power Consumption of Various Control Systems

Reprinted by Permission: *Energy Engineering, Vol.* 86, *No.3*: "Profit Improvement with Variable Frequency Drives," Scott A Moses*, Wayne C. Turner, Ph.D., P.E., CEM, Jorge B. Wong, Mark R. Duffer, Oklahoma State University, Stillwater, OK.
*This material is partially based upon work supported under a National Science Foundation Graduate Fellowship.

Any opinions, findings, conclusions or recommendations expressed in this publication are those of the authors and do not necessarily reflect the views of the National Science Foundation.

is determined by its frequency, although a constant ratio of voltage to frequency must be maintained for the motor to produce full torque. Thus, a VFD responds to a change in load by adjusting the frequency and voltage of the power supplied to the motor, which changes the motor's speed.

Design Features
Sizing

A centrifugal fan or pump motor is sized to meet maximum load, but usually this load occurs only a few hours a year. Often, a fan motor is oversized by a factor of two or more, partially due to conservative design practices and also to overcome the starting inertia of a fan.

During off-peak periods, the load fluctuates, and either' dampers or throttling valves control the flow. This can be compared to driving a car with your foot on the accelerator and the brake at the same time, which is wasteful. However, the amount of energy wasted is substantial, since the power requirement of a centrifugal fan or pump varies with the *cube* of the speed, while the flow varies *directly* with the speed. Thus, power consumption drops drastically when speed is reduced. For example, reducing fan speed 20 percent reduces air flow 20 percent, but the power consumed is reduced by nearly 50 percent.

Air-handling Units and Centrifugal Fans

Air-handling units consume a major fraction of the energy required to heat and cool a space. Originally, all air handlers were one-speed, constant volume units. Since a fan rarely needs to run at full speed in order to meet space cooling requirements, different control strategies were developed to reduce the fan horsepower during off-peak periods. Often, constant volume systems have been replaced with more efficient Variable Air Volume (VAV) systems. To modulate output flow, two traditional methods of controlling air flow in a VAV system have been used: Variable Inlet Vane (VIV) control and Outlet Damper (OD) control. Both of these methods reduce the fan input power. However, neither VIV nor OD control reduce fan speed, which would result in maximum savings.

Pumps

Similarly, pumping systems must be designed for maximum needed flow rates. Frequently, either extra flows bypass the use point and return to a reservoir, or throttling valves reduce the flow, or pressure builds to the point where centrifugal pumps are virtually spinning with very little volume delivered. Sometimes, extra flows are delivered to the use point at design volume regardless of need.

Alternatively, flows can be controlled to match need, and a pressure sensor at the pump outlet can be used to send signals to a VFD. The VFD then controls the speed of the motor to deliver a constant pressure, rather than a constant volume. Energy savings curves for pumps have the same shape as those for fans, so energy savings can be large and payback rapid.

Other savings include maintenance savings and reduced downtime. The life of belts, pulleys, bearings, and motors will be lengthened. The VFD will require no routine maintenance. Bypassing the VFD drive to allow the motor to operate at full speed is simple, so there is no downtime if the VFD does require repair.

Energy savings will be primarily determined by the load profile for the process. A VFD will be attractive economically if the load profile varies widely and low to medium loading occurs frequently. As noted earlier, constant volume pumping or air-blowing systems are excellent candidates when the actual load varies but the motor is run at full speed. For example, a typical duty cycle of an air conditioning system with VAV control is shown in Figure 13-5.

Secondary considerations which will improve the attractiveness of a VFD are: long operating hours, higher energy costs for electrical consumption, and a medium-size motor. A VFD saves kilowatt-hours. It usually

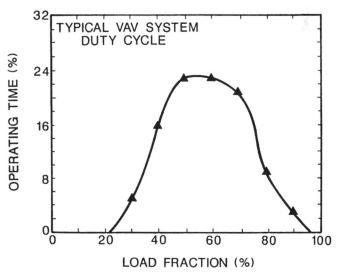

Figure 13-5. Typical VAV System Duty Cycle

will not save on demand, unless the motor is oversized and/or the period of *low* motor loading coincides with the system peak period. Medium-size VFDs tend to be most economical. Small drives may have a small savings-to-investment ratio, and very large VFD drives tend to be prohibitively expensive.

Candidate Applications

Any fan or centrifugal pumping system with a varying load makes a good candidate. Following are some typical ones:

- Air-handling units. Conversion from a constant volume, variable temperature system to VFD control with constant temperature is especially attractive. Even more efficient VAV systems with a high percent throttling have energy saving potential. A fan in a typical VAV system runs at 80 percent speed or less 90-95 percent of the time, which meets the widely-varying, low load requirement. (See Figure 13-5 for a typical duty cycle profile.)
- Cooling tower fans.
- Pumping applications include municipal water systems, chemical/ petrochemical industries, pulp and paper industry, food industries, machine coolant and pumping systems.
- Mixers, conveyors, packaging and bottling machinery, dryers, crushers, grinders, and extenders.

Step-by-step Economic Evaluation

Step 1: Collect Data

To calculate the annual savings generated by installing a VFD on an air-handling unit, obtain the following data:
- Motor size (hp)
- Total annual operating hours (hrs/yr)
- Incremental cost of electricity ($/kWh)
- Type of system (constant volume, outlet damper, variable inlet vane, or eddy-current drive)

Step 2: Calculate Annual Savings

Table 13-1 gives the consumption savings based on a typical duty cycle. If the system peak and the period of low motor loading coincide, demand savings should be added to the total annual dollar savings. The method for filling in the blanks is given in the notes.

Table 13-1. Values for Calculating Annual Savings

Load Ratio 1	Power Input Ratio (Old) 2	Power Input Ratio (VFD) 3	Duty Cycle Fraction 4	KWh Saved 5	Dollar Savings 6
0.20	—•—	0.09	0.00	———	———
0.30	—•—	0.11	0.05	———	———
0.40	—•—	0.14	0.16	———	———
0.50	—•—	0.20	0.23	———	———
0.60	—•—	0.29	0.23	———	———
0.70	—•—	0.43	0.20	———	———
0.80	—•—	0.62	0.09	———	———
0.90	—•—	0.85	0.03	———	———
1.00	—•—	1.16	0.01	———	———

Total Annual Electricity Savings ...._____kWh
Total Annual Dollar Savings...$_____

Column 2: Obtain the values from Table 13-2
Column 4: Source: Graham Co., Milwaukee, WI. The values in Table 13-1 are for a standard duty cycle and should be used unless a typical duty for' the entire year has been determined. Actual fractions from measurements will give much more accurate results. Simply plug the new fractions into Column 4.
Column 5: For each Load Ratio, calculate kWh Saved = (0.7457)(hp)(hrs)(Column 2 - Column 3)(Column 4) where
Hp = Rated Motor Size, hp
Hrs = Hours of Operation, hrs/yr
Column 6: For each Load Ratio, calculate Dollar Savings = (Column 5)(COE) where
COE = Cost of Electricity, $/kWh

Table 13-2. Power Input Ratios (Including Motor and System Losses)

	Power Input Ratio			
Load Ratio	Constant Volume*	Outlet Damper	Inlet Vane	Eddy Current
0.20	1.10	0.64	0.57	0.04
0.30	1.10	0.73	0.67	0.09
0.40	1.10	0.82	0.71	0.16
0.50	1.10	0.90	0.72	0.26
0.60	1.10	0.97	0.75	0.39
0.70	1.10	1.05	0.80	0.54
0.80	1.10	1.11	0.88	0.71
0.90	1.10	1.17	0.99	0.91
1.00	1.10	1.22	1.20	1.12

*These ratios are specifically applicable to single zone, constant volume systems.

Step 3: *Determine Installation Cost*

Based on the voltage constraints of each installation, up to five possible configurations are possible (see Figure 13-6). Since the voltage will not affect the annual savings, each feasible configuration will be considered and the least expensive one selected. The five possible configurations are shown in Figure 13-6 and described as follows:

1. 460V line voltage, 460V VFD, 460V existing motor
2. 460V line voltage, 460V VFD, 460V new motor (to replace a 230V motor)
3. 230V line voltage, 460V VFD, 230/460V dual-voltage motor, transformer
4. 230V line voltage, 230V VFD, 230V existing motor
5. 230V line voltage, 460V VFD, 460V new motor, transformer

Table 13-3 will assist in selecting the best configuration option that is compatible with your application.

Step 4: *Evaluate the Investment*

Typically, a simple payback will be calculated to determine the economic merit of an investment. This method is a rough approximation, since it neglects the time value of money. Typical paybacks of 2-3 years or less underestimate the true period until breakeven.

Net Present Value Method

In addition to calculating a simple payback, the net present value

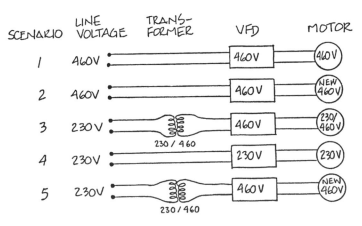

Figure 13-6. Typical VAV System Duty Cycle

Table 13-3. Scenario Selection Matrix

Available Line Voltage Only	230V Only		230V and 460V		460V	
Motor Voltage	230V	230V/- 460V	230V	230V/- 460V	230V	230V/- 460V
Best Scenario	4	3	2 or 4	1	1	1

(NPV) should be obtained in the following steps:

(1) You should determine the annual savings and the installation cost of a VFD in the above analysis.

(2) Decide on the number of years, N, over which you want to spread your investment. A good choice would be 4 years. The smaller the N, the harder it is to justify an investment.

(3) Predict the interest rate over the next N years. This is your minimum attractive rate of return (MARR).

(4) Determine the present worth factor from a set of interest tables or use the formula below:

$$\text{Present Worth Factor} = \frac{(1 + i)^N - 1}{(i)(1 + I)^N}$$

where

i = the decimal effective interest rate per period, such as 12 percent per year, which would be 0.12

N = the number of periods of investment, such as 4 years.

(5) Compute the net present value (NPV) as follows:

NPV = (total annual savings) (present worth factor) – (Investment Cost).

The NPV will be a dollar amount equivalent to a lump sum of the savings you can expect over the next N years.

Decision Criteria

If the NPV is positive, the investment will result in a net gain over the period of N years. In other words, install the VFD. If the NPV is negative, it is not an attractive investment.

This is a conservative analysis because the annual savings calculated above are strictly from energy cost reductions. They do not include maintenance savings (such as the elimination of inlet vane or outlet damper maintenance), electrical demand savings, or savings from reduced downtime of the new system. The advantages of a VFD, which might be enough to warrant an investment on its own, have not been quantified. Also, the salvage value of the equipment has been neglected.

The purpose of this analysis is to determine the economic merit of a VFD for a certain application. If a VFD merits further consideration, the application should be evaluated by a qualified consultant. Some of the factors to consider include operating environment, possible overloads, maximum current that the drive must handle, and optional components which might be needed.

An Office Building

An office building in Oklahoma has two air-handling units. The larger unit is 50 hp, is controlled by variable inlet vanes, and operates 6500 hrs/yr. It is a 440V unit. The smaller unit is a 20 hp constant volume unit and operates 4000 hrs/yr. The present motor is 230V, but both 230V and 440V line voltages are available. The electricity costs are $0.16/kWh.

Calculation of Annual Savings

The values in Column 2 of Table 13-4 are obtained from Table 13-2. Columns 5 and 6 are calculated using the previous formulas.

The total savings for this example is about $25k and 155,000 kWh of electricity every year.

Determination of Installation Cost

Using Table 13-3, the least expensive configuration is determined for the large air handler, which will be Scenario 1. For the small air handler, the best configuration is either Scenario 2 or 4, depending on the specific size of the motor.

VFD on Dust Collectors

A stainless steel foundry in Milwaukee, WI, has two cyclone dust collectors, both of which have many pieces of equipment with varying loads connected to them. One, a 100 hp unit, has 20 pieces of equipment attached to it. The other, a 50 hp cyclone unit, has 15 pieces of equipment

Table 13-4. Annual Savings for a Large Air Handler

1 Load Ratio	2 Power Input Ratio (Old)	3 Power Input Ratio (VFD)	4 Duty Cycle Fraction	5 kWh Saved
0.20	0.57	0.09	0.00	0
0.30	0.67	0.11	0.05	6786
0.40	0.71	0.14	0.16	22103
0.50	0.72	0.20	0.23	28985
0.60	0.75	0.29	0.23	25641
0.70	0.80	0.43	0.20	17934
0.80	0.88	0.62	0.09	5671
0.90	0.99	0.85	0.03	1018
1.00	1.20	1.16	0.01	97

Total Annual Electricity Savings....................108,235 kWh

Table 13-5. Annual Savings for a Small Air Handler

1 Load Ratio	2 Power Input Ratio (Old)	3 Power Input Ratio (VFD)	4 Duty Cycle Fraction	5 kWh Saved
0.20	1.10	0.09	0.00	0
0.30	1.10	0.11	0.05	2953
0.40	1.10	0.14	0.16	9163
0.50	1.10	0.20	0.23	12349
0.60	1.10	0.29	0.23	11114
0.70	1.10	0.43	0.20	7994
0.80	1.10	0.62	0.09	2577
0.90	1.10	0.85	0.03	447
1.00	1.10	1.16	0.01	–36

Total Annual Electricity Savings....................46,562 kWh

connected to it. Both are 460V cyclone motors and operate 7,500 hrs/yr.

The cyclone systems are set up so that when a piece of equipment is shut off, a damper closes on that particular branch of the cyclone ducting. In that manner, the system acts like a variable inlet vane system. In the following example the typical VAV system duty cycle is applied, with electricity costs of $0. 32/kWh.

Calculation of Annual Savings
The values in Column 2 of Tables 13-6 and 7 are obtained from Table 13-2 for inlet vanes. As before, Columns 5 and 6 are calculated using the formulas presented earlier.

The total savings for both investments are $12,031 and 374,809 kWh annually.

Determination of Installation Cost
Using Table 13-3, we determine that the least expensive configuration for both dust collectors will be Scenario 1. For the 100 hp cyclone, the installation cost is determined to be $9,572. For the 50 hp cyclone, the installation cost is $5,663. The total investment cost will be $15,235.

Evaluation of the Investment
Once again, the simple payback will be given, but the decision will be based on the net present value at a 12 percent interest rate and a 4-year investment life.
Simple Payback = ($15,235)/($12,031/yr) = 1.27 years
Net Present Value = ($12,031)(3.0373) – ($15,235) = $21,306

Since the NPV is positive, this is a very good investment. However, it is also important to check the minimum velocity required in a dust collection system, which is necessary to avoid settling. That requirement may affect the economics of the project.

Table 13-6. Annual Savings for 100 hp Dust Collector

1 Load Ratio	2 Power Input Ratio (Old)	3 Power Input Ratio (VFD)	4 Duty Cycle kWh Fraction	5 kWh Saved	6 Dollar Savings
0.20	0.57	0.09	0.00	0	0
0.30	0.67	0.11	0.05	15666	503
0.40	0.71	0.14	0.16	51026	1638
0.50	0.72	0.20	0.23	66916	2148
0.60	0.75	0.29	0.23	59195	1900
0.70	0.80	0.43	0.20	41403	1329
0.80	0.88	0.62	0.09	13092	420
0.90	0.99	0.85	0.03	2350	75
1.00	1.20	1.16	0.01	224	7

Total Annual Electricity Savings...............249,872 kWh
Total Annual Dollar Savings.....................$8,020/yr

Table 13-7. Annual Savings for a 50 hp Dust Collector

1	2	3	4	5	6
Load Ratio	Power Input Ratio (Old)	Power Input Ratio (VFD)	Duty Cycle kWh Fraction	kWh Saved	Dollar Savings
0.20	0.57	0.09	0.00	0	0
0.30	0.67	0.11	0.05	7833	251
0.40	0.71	0.14	0.16	25513	819
0.50	0.72	0.20	0.23	33458	1074
0.60	0.75	0.29	0.23	29598	950
0.70	0.80	0.43	0.20	20702	665
0.80	0.88	0.62	0.09	6546	210
0.90	0.99	0.85	0.03	1175	38
1.00	1.20	1.16	0.01	112	4

Total Annual Electricity Savings...............124,937 kWh
Total Annual Dollar Savings.....................$4,011/yr

VFD Attributes

These are just a few of the examples to show the savings calculations. The following summarizes the advantages of VFDs.

Efficiency

VFDs operate at about 85 to 95 percent efficiency over their entire speed range. For example, consider an air-handling unit operating at 50 percent of full load. A system controlled by a VFD would consume 85 percent less energy than a constant volume system, 80 percent less than a system with outlet damper control, and 75 percent less energy than a system with variable inlet vane control.

Compensation for Oversized Designs

A VFD will effectively size the motor to the load, regardless of motor horsepower. This could be good on a retrofit, although a new and smaller motor might be more cost effective.

High Power Factor

The power factor of a VFD is typically about 0.95, which may generate further dollar savings on the electric bill, due to a reduction in excess kVAR (kilovolt ampere reactants).

Some utilities penalize heavily for low power factors, leading to significant dollar savings when the power factor is improved.

Easy to Retrofit

No mechanical fabrication is required, and the VFD can be remotely mounted. Installation takes only a few hours. VFDs operate safely in hazardous atmospheres, and an existing motor can be used.

Designed to Work with Standard AC Motors

Interface Easily with Computers

A VFD can be connected easily to a computer energy management system. The VPD will not cause electronic interference with telecommunications or computer equipment when the VFD is properly outfitted.

Have Soft Starting Capability

When a normal motor is started, the surge of electric current that moves the motor out of its stationary position is about six times the ordinary current. The current in-rush produces stress on the motor windings and other components. However, the acceleration time on a VFD can be adjusted from instantaneous to several minutes to reduce the stress.

Improve Equipment Life

Soft starting and the lower average operating current of a VFD extend equipment life. The reduced current and voltage drop associated with soft starting will reduce motor heating. However, equipment life is improved only if the power is properly conditioned; if not, negative effects can occur, such as overheating.

Re-acceleration and Rotating Motor Restart Capability

If the power should be briefly interrupted, the VFD will resume operation immediately, whereas fixed speed devices would not. The drive will restart the motor at the speed at which it is rotating.

Not Just an Energy-saving Device

VFDs are often used in speed control applications because of their accuracy and their ability to control a number of synchronized motors. They can also be used to control a process more uniformly, and soft starting can be used to avoid "jerk" on conveyors.

AC VARIABLE SPEED DRIVES / INVERTERS EXAMPLES*

Figure 13-7. AutomationDirect AC variable speed drives for speed control applications. *(Used with permission of AutomationDirect)*

GSI Series Mini-drives (Automation Direct)

The GSI series of AC drives have V/Hz control with advanced application features. These drives can be configured using the built-in digital keypad (which also allows you to set the drive speed, start and stop, and monitor specific parameters) or with the standard RS485 serial communications port. Standard GS1 features include one analog input, four programmable digital inputs, and one programmable normally open relay output.

GS2 Series Micro-drives (Automation Direct)

The GS2 series of AC drives offers all of the features of our GS1 drives plus dynamic braking, Pill and a removable keypad. The drive can be configured using the built-in digital keypad or the standard RS485 serial communications port. The standard keypad allows you to configure the drive, set the speed, start and stop the drive, and monitor specific parameters during operation. Each GS2 AC drive features one analog and six programmable digital inputs, as well as one analog and two programmable relay outputs.

GS3 Series DURApulse AC Drives (Automation Direct)

The DURApulse series of sensorless vector drives, built on the simplicity and flexibility of the GS1 and GS2 serial drives, are available in 1 to 100 Hp models and offer sensorless vector and closed loop control autotuning, 175% starting torque, and 150% rate current for one minute.

The DURApulse series of AC drives offers all of the features of our

---

*From Automation Direct 2008 Catalog.

GS2 series drives, including dynamic braking PID, removable keypad, and RS485 MODBUS communication. The DURApulse drives also offer sensorless vector control, with the option of encoder feedback for enhanced speed control.

The standard smart keypad (a.k.a. HIM or human interface module) is designed with defaults for North American customers and allows you to configure the drive, set the speed, start and stop the drive, and monitor critical parameters for your application. In addition, this keypad has internal memory that allows four complete programs to be stored and transferred to any DURApulse drive. The DURApulse series offers three analog inputs, eleven digital inputs, and one SPDT relay output.

Features of Durapulse AC Drives:
- Simple Volts/Hertz control
- Sensorless vector control with autotune
- Sensorless vector control with optional encoder feedback card, for better speed control
- Variable carrier frequency, depending on model. IGBT technology
- 175% starting torque
- 150% rated current for one minute
- Internal dynamic braking circuit for models under 20 Hp
- Automatic torque and slip compensation
- Programmable jog speed
- Removable smart keypad with parameter upload/download
- HIM keypad with memory to store up to four programs of any DURApulse drive Three analog inputs and one analog output
- Eleven digital inputs
- Three digital and one SPDT relay output
- One analog output
- RS485 MODBUS communications
- Ethernet communication optional
- UL/CE listed
- Optional software package with full programmability, trending, and application setup

Typical Applications of DURApulse AC Drives
Some of the typical applications for which DURApulse AC drives are used include: conveyors, fans, pumps, compressors, HVAC, material handling, mixing shop tools, extruding, and grinding.

Typical Specifications of DURA pulse AC Drives
200-240 VAC, 380-480 V AC, HP → 100 HP

References
"DDC-Type Control Applied to Motors," *Energy User News.*

Doll, Thomas R. "Making the Proper Choice of Adjustable-Speed Drives," *Chemical Engineering.*

Fugill, Richard W. and Philibert, Claude. "What You Should Know About Adjustable-Speed Drives," *Electrical Construction and Maintenance.*

Helmick, C.G. "Applying the Adjustable-Frequency Drive," *Electrical Construction and Maintenance.*

Automation Direct 2008 Catalog.

Chapter 14

Field Measurements and Determination of Electric Motor Efficiency*

Conservation of electricity in electric motors and the systems they drive has been a subject of increasing concern recently. Electric motor manufacturers have introduced more efficient motors in response to the sharply increasing price of electricity. A number of papers have appeared about the efficiency of electric motors and the procedure for selection of an appropriate motor to perform a particular task.[1,2,3,4,5] Many users of motors now include efficiency as one of the parameters for evaluation of motors for purchase, including independent tests on motors.[2,3]

Almost all of these papers address the issue of new motors and are useful when the decision to purchase a new motor has already been made. However, there is little consideration for the millions of motors already operating in the field. Some of these motors may not be performing their task efficiently, and a cost effective alternative toward better performance may exist. The reasons for inefficiency or otherwise poor performance may be many. For example, there may be an inherently inefficient motor, an improper match to load, a poor rewind job (if rewound), or voltage imbalance.

Evaluating Operating Condition

The first step towards determining a future course of action is the evaluation of the present operating condition of the motor. This paper reports work towards several methods to evaluate operating conditions:

- Determining a suitable procedure for field measurements, including a package of instrumentation and data acquisition hardware to facilitate large-scale field testing of the motors

*Reprinted with special permission; K.K. Lobodovsky, R. Ganeriwal, and A. Gupta, PG&E Company, *Energy Engineering*, Vol. 86, No. 3.

- Developing a computer program to process the motor nameplate and the measured values to calculate the motor performance map and obtain an economic evaluation of replacement/retrofit/do-nothing alternatives

- Setting up a database to simplify field motor evaluation in the future and provide more accurate information for electricity demand forecasting and conservation program assessment activities

Objectives and Applications

Table 14-1 summarizes three major objectives and the applications of this program. The overall motivation for promoting this program, along with the explanation of the specific objectives and applications, is to achieve economically optimal efficiency that is beneficial for all parties involved in the program, including:

- The plant operators—obtaining more competitive production costs;*
- The utility—avoiding new capacity by reducing wasteful usage;
- The state—continuing lower electricity rates and enhancing industrial productivity.

The field evaluation of motors provides a means for operators to make informed decisions on the currently operating electric motors. More specifically, the objectives of the program can be divided into three areas of interest:

- Motor performance from field measurements for determination of applicable alternatives.
- Economic evaluation of alternatives to improve motor efficiency.
- Long-term database development for planning utility conservation programs.

The program provides an estimate of the motor efficiency from field measurements at the operating condition(s), along with the complete performance map or the operating characteristics throughout the no load to

---

*It may be noted that all the results reported in this paper are based on tests conducted in industrial plants. However, the insights and experience are equally applicable in nonindustrial applications.

Table 14-1. Objectives and Applications of Field Motor Testing Program

OBJECTIVES:		APPLICATIONS:	
2.1	To establish motor performance map under field conditions	2.1	To determine applicability of alternatives (e.g., reduce the size of the motor, NOLA device for under-loaded motor, etc.)
2.1.1	Complete measurements: Accurate estimates of motor efficiency	2.1.1	provide information for decision-maker to retrofit, replace or do nothing on the motor
2.1.2	Partial measurements: (e.g., due to field conditions, only the load test feasible) Probabilistic or no estimate or motor efficiency.	2.1.2	provide information for: —understanding more complete measurements —decision to retrofit, replace, or do nothing —assistance in future maintenance and replacement/rewind decisions
2.2	To evaluate economic viability of alternatives to improve the efficiency of motor and motor-driven systems	2.2.1	To determine the cost-effectiveness of retrofit and replacement alternatives for the plant operator
		2.2.2	To determine the cost-effectiveness of incentive and other conservation programs initiated by utility companies or public agencies
2.3	To build a data-base from field evaluations and manufacturers'/-rewinders' data	2.3.1	To establish probabilistic motor evaluation procedures where site specific factors prohibit complete measurements to assist the operator in making the best decision
		2.3.2	To assist planning for future energy demands and conservation program evaluation at the utility service area level on the basis of actual efficiencies and the likely response of the users.

rated service duty range. This in turn allows the determination of alternatives that are likely to be applicable for the actual conditions under which the motor is operating. For example, if the motor is found to be under-loaded, downsizing and NOLA devices can be considered, in addition to the replacement by a new more efficient motor.

The procedure chosen for estimating motor performance is IEEE 112 Method E/F. The measurements required for this method are load and no load voltages, amperes, power factor (or power), shaft RPM and the stator resistance. It should be noted that for an original (not rewound) motor, if the manufacturer's detailed performance specifications are available, much simpler measurements will meet the objectives. But, this is seldom found to be true in field conditions.

Measurement Difficulties

Our field experience shows that it is not always easy to obtain all the requisite measurements. This is especially true of the no load measurements, which require uncoupling the motor. If these measurements can be taken, the complete test provides an accurate estimate of motor efficiency at the operating load point. This further permits calculation of relevant motor parameters to allow a complete characterization of the motor performance at any load point. The performance characteristics include efficiency, power factor, line current, power input, motor speed, and any other outcome of interest. In Table 14-1, this is shown under 2.1.1. In such cases, the program provides complete information to enable the operator to evaluate the decision to replace, retrofit or do nothing on that particular motor.

On the other hand, partial measurements in many instances can provide some clues about the motor. For example, unusually high no load power input may be sufficient to justify replacement. Thus, a second objective, shown under 2.1.2 in Table 14-1, is to determine the conditions where even the partial field measurements can provide sufficient information for either a final decision (to replace, or not replace, or retrofit) or an indication as to whether a full test is warranted.

Economics of Alternatives

The second part of the program involves analysis of the economics of various alternatives for the motor conditions determined in 2.1 in Table 14-1. In addition to the motor efficiency and the nameplate rating obtained earlier, the operator has to supply the total operating hours in a year, the duty cycle (if it is not a constant on-off application), and the load associated with each point in the duty cycle. The electricity prices over time are taken from the utility, state or operator's own projections.

The costs of implementing alternatives are based on vendor information. The economics of alternatives are calculated in terms of simple payback and a complete internal rate of return (accounting for taxes, depreciation and any tax or other incentives provided to a technology option). This chapter will briefly discuss the methodology and results of 29 motor tests, in addition to determining the economics of retrofit, replacement, and do-nothing alternatives for the operator.

This information will be used to evaluate the cost-effectiveness of any programs initiated by the utility company or public agencies. Such programs can provide several alternatives, from direct subsidies to information dissemination and audit services.

The final objectives consist of developing a long-term database of field test results, manufacturers'/rewinders' data, and the follow-up results of the actions by the plant operators. This is likely to serve several purposes. It is our hope that with sufficient field experience in conjunction with the cooperation from the motor industry (both manufacturers and the repair/rewind industry), general trends and rules of thumb can be developed. This allows inference of motor behavior even with incomplete tests. This goal is cited in Table 14-1 at 2.1.2. In addition, the database on results of *actual* efficiencies of field equipment provides planning agencies with the ability to forecast future demand from this important electricity-using sector. A U.S. Department of Energy study estimates the motor power consumption to be in the order of 80 percent of the industrial and commercial sector's electricity demand.[8] Lastly, the follow-up, whether it involves implementation or nonimplementation of the recommendations based on this program, is likely to provide useful information for the program administrators, as well as the vendors of various types of equipment.

Test Procedures and Evaluation Programs
Test Procedure

The procedure adapted for motor testing was in accordance with the IEEE Standard 112-1978, a combination of methods E and F as outlined in the standard.[6] The standard is quite detailed, and further explanations appear in the IEEE Transactions on Industry Applications.[1] This involves the following tests:

- No load test. The motor is to be completely decoupled from the load and turned on. The voltage, current, and power readings are to be taken under the no load condition.
- Stator resistance. The stator resistance of the motor is to be calculated

by measuring the line-to-line resistance. Temperature corrections are to be applied to all readings.

- Load test. Voltage, current and power input readings are to be taken under actual load condition. Motor speed is also to be measured under the actual load condition.

Some of the important assumptions which were made in our calculations are:

(1)   The locked rotor test is not performed. Instead, the ratio of the stator reactance to the rotor reactance is assumed according to the NEMA design of the motor. The ratios are given in the IEEE test procedure.

(2)   The stator temperature rise is assumed on the basis of the insulation class, as outlined in the IEEE test procedure.

(3)   No measurements are taken to determine the stray load loss. In accordance with ANSI C50.41,[7] this loss is assumed to be 1.2 percent of the rated output under full load conditions. For partial loads, this loss is assumed to be directly proportional to the square of the load.

(4)   The reduced voltage test under no load condition is not performed. The friction and windage losses are thus coupled with the core losses, and only the sum of these two losses can be calculated from the no load test. This result is a small decrease in the calculated rotor resistance loss. The net effect of the absence of this test on efficiency calculations is negligible.

(5)   All electrical measurements are made at the motor starter terminals. Corrections are applied to account for the resistance of the cable from the starter to the motor, as well as the associated voltage drop and power loss.

Speed is measured under no load conditions to ensure that the motor is indeed unloaded. This should result in a speed almost exactly equal to synchronous speed. Supply voltage values and the line frequency are monitored for any significant voltage imbalance and departure from 60 Hz frequency. IEEE procedure specifies the acceptable tolerance to be ±5 percent imbalance.

Instrumentation

After initial experimentation with various instruments, a DRANETZ 808 was chosen for voltage, current, and power measurements, whereas shaft rpms were measured with a digital strobe-scope. The experience leading to these choices is briefly described in the next section, and the goals for improvements for the future are covered in the following section.

The Computer Program

A computer program was written to process the information collected from the tests. The program gives very detailed information about the motor. Not only does it provide us with complete information about the motor behavior under the existing operating conditions, but it also predicts the motor behavior under different operating conditions (voltage, and load point, etc.). The program also calculates the financial attractiveness of alternatives.

The memory and processing time requirements of the program are reasonably small. The program has been checked against other programs written for a similar purpose with excellent agreement.[1]

Evaluation of Financial Attractiveness

Two basic parameters are used to evaluate the financial attractiveness of the alternatives—simple payback and internal rate of return.

The yearly energy savings can be calculated by summing up the energy savings at all the different loads at which the motor operates over the year.

$$
\begin{aligned}
\text{Energy savings \$/yr} \\
\text{(at a particular load)}
\end{aligned}
\quad
\begin{aligned}
&= \text{Motor size in HP} \times .746 \\
&\times \text{load (fraction of rated size)} \\
&\times \text{Hours/Year of operation at this load} \\
&\times \left[\frac{1}{E_1} - \frac{1}{E_2} \right]
\end{aligned}
\qquad (14\text{-}1)
$$

Where E_1: Efficiency of the existing motor at this load
$\quad\quad$ E_2: Efficiency of the alternative at this load

$$
\text{Simply payback} = \frac{\text{Capital cost of the project}}{\text{Total energy savings per yr}} \qquad (14\text{-}2)
$$

Calculating the internal rate of return (IRR) is a somewhat more complicated calculation, but it is a more accurate measure of the financial attractiveness of an alternative.

An after tax, discounted cash flow analysis was used to calculate the IRR. Real (or constant) dollars were used in the analysis, so the result gives the real rate of return on the capital after adjustment for inflation. Such an analysis accounts for depreciation, investment tax credit, electricity price escalation rates, tax rates, and rewind costs. The formulas for this calculation are a little complex, but they are a part of standard accounting practices.

Field Experience

Before presenting the results, some general remarks about our experiences are in order.

The test results presented in this chapter are from the first phase of a two-phase project where one of the aims of the small pilot testing program was to assess the various steps involved in carrying out the overall effort. The steps involved not only the testing of instrumentation, where some ideas for improvements have emerged and are being implemented for future testing, but also the cooperation of the plant operators and the testing personnel. All in all, it is an arduous undertaking and in the long run would require a basic interest and commitment of the plant personnel.

After some experimentation with different instrument configurations, it was decided that the goal should be a package that provides simultaneous hard copies of readings, hence minimizing errors due to changes in loads. A digital read-out strobe-scope was determined to be more suitable than a magnetic pickup device for obtaining the motor rpm. The DC resistance readings were taken with a separate multimeter. All instruments were calibrated in the lab before testing began.

In the long run, the DC resistance readings and signals from the rpm measuring device will be inputted and recorded along with the electrical readings. This will eliminate the possibility of the rpm readings being out of synchronization with *volts/amps/power* readings. This will add the ability to establish the efficiency of the motor under time-varying conditions.

All the improvements, when incorporated, should result in a convenient and efficient procedure.

Site Tests

The first plan of the program involved field measurements at four industrial plants in Northern California. A total of 60 motors were tested, out of which both load and no load tests were completed on 29 motors.

The four plants where on-site motor tests were conducted varied widely in every respect. The processes included glass, paper, sand and gravel, and anode grade coke making operations. The functions and states of the motor were equally divergent.

The measurements in one plant were made while the process was operating. In several instances this involved minor interference with the process, and the no load tests had to be rushed because of the extreme limitation on the time that the motor could be kept decoupled from the equipment. In two plants, the no load measurements were performed when the production line was down. One plant ran one shift only, the no-load tests were performed after the shift was out, whereas the no-load testing in the other plant had to be scheduled when the process line came down for biweekly maintenance.

In the last plant, where the process runs continuously, a total of 17 motors were tested, but complete load/no load tests were feasible only on three motors. The rest were completed when the plant was down for a major revamping operation.

Thus, the complete IEEE 112 method E/F load and no-load tests are quite constraining in the industrial operating environment. Of course, if the tests are carried out in close cooperation with (or by) the plant operators, the scheduling and other coordination logistics are reduced.

In our tests, there were two persons at the testing end—one setting up and taking the electrical readings while the other took the motor nameplate data and the rpm reading. Generally, it took two maintenance personnel from the plant to uncouple and put the motors back together. On the average, a complete test took approximately a half hour. Most of the motors selected for the present set of tests were belt driven.

Table 14-2 presents results of the economics in replacing 29 motors with higher efficiency motors. The data on costs and nominal efficiencies of the replacement motor were based on published figures from a leading motor manufacturer. Even though the plants tested did not operate the same number of hours each year, a figure of 8,200 hours/year was used for consistency. The electricity price was assumed to be 7.0¢/kWh, with an escalation rate of 2 percent per year in real (constant) dollars. For purposes of calculating the internal rate of return (IRR), a tax life of 10 years, with a straight line depreciation and 10 percent investment tax credit were assumed. The federal and local tax rate was taken to be 55 percent. All recommendations are based on a 20 percent minimum IRR.

Downsizing Motors

A large number of motors are operating at loads far below their rated size. On an average, they are operating at 60 percent of the rated load. It is

Table 14-2. Results of Field Testing of Industrial Motors

No.	Size (hp)	Application	Load (% Rated Size)	Efficiency At This Load %	Power Factor At This Load %	Simple Payback (years)	IRR* (%)	Recommendation
1	10	Conveyor	22.9	74.7	40.8	1.4	41.5	Downsize
2	10	Conveyor	23.4	76.9	42.0	1.6	35.9	Downsize
3	10	Screen	78.3	86.8	79.2	2.9	21.8	Replace
4	15	Conveyor	56.8	85.4	61.3	2.0	29.8	Downsize
5	15	Screen	64.9	89.0	71.0	4.2	15.4	Keep this motor
6	15	Pump	93.6	90.4	87.3	5.4	12.1	Keep this motor
7	20	Conveyor	14.8	63.1	49.3	0.6	90.1	Downsize
8	20	Pump	53.0	86.6	64.0	2.4	25.6	Downsize
9	20	Vacuum Pump	101.5	88.1	85.3	1.9	31.8	Replace
10	20	Compressor	112.6	88.1	82.3	1.7	34.3	Replace
11	20	Blower	65.9	88.5	80.4	2.9	21.5	Downsize
12	20	Blower	70.5	89.0	83.5	3.2	20.1	Replace (Marginal)
13	25	Screen	45.9	85.4	57.0	1.8	32.8	Downsize
14	25	Pump	75.7	92.1	76.0	6.1	10.6	Keep this motor
15	30	Screen	42.4	86.7	67.2	2.0	29.5	Downsize
16	30	Screen	53.2	87.0	77.4	1.9	31.3	Downsize
17	30	Screen	34.3	89.5	60.3	5.9	11.0	Keep this motor
18	30	Pump	87.8	90.9	77.8	2.9	21.8	Replace
19	40	Screen	57.9	85.0	81.0	1.1	50.8	Downsize
20	40	Mixer	65.1	85.3	82.0	1.1	51.4	Downsize
21	40	Screen	48.7	87.1	74.4	1.6	35.9	Downsize
22	40	Screen	48.3	87.8	72.7	1.9	31.9	Downsize
23	40	Pump	75.0	92.0	90.9	3.7	17.5	Keep this motor
24	50	Blower	57.2	89.8	70.7	2.0	29.6	Downsize
25	60	Pump	65.6	90.1	81.1	2.1	28.6	Downsize
26	75	Blower	22.5	81.7	57.8	1.1	52.0	Downsize
27	150	Pump	72.8	93.6	86.4	6.8	9.2	Keep this motor
28	250	Scrubber	55.1	87.9	80.4	1.4	41.9	Downsize
29	250	Scrubber	67.4	95.8	91.5	59.1	−6.5	Keep this motor

*After tax, real (adjusted for inflation), Internal Rate of Return

not surprising that the recommendation for 17 out of 29 of the tested motors is replacement with a smaller motor. However, downsizing may not be a feasible option in applications where the process could require extra power under unusual circumstances. Further, some motors may have poor access. A change in frame size, in some instances, can impose additional costs. Even without downsizing, 15 of these 17 motors can be economically replaced, though the economics of replacement is not as good. The possibility of the applicability of a NOLA-type power factor controller on some of the heavily underloaded motors is being investigated.

It should be pointed out that we have tried to be conservative at each step in our financial calculations. The actual rate of return on the investment will probably be higher, due to reduced losses in power supply cables and transformers, since efficient motors draw less current. A better power factor and longer motor life will further reduce costs. However, more efficient motors usually operate at a slightly higher speed, resulting in lower rotor resistance losses. This could result in increased power consumption in certain applications, like blowers and centrifugal pumps, where the power needed is dependent on speed. This fact has been generally neglected and in our opinion requires serious consideration when evaluating higher efficiency (lower slip) motors.

Seven motors were found to be performing their task quite efficiently and required no further action. Another five, though properly sized, would be candidates for replacement with favorable economics.

Based on this research, Esterline Angus Co. of Indianapolis, IN, actually designed the instrument for motor testing and used the software PG&E developed for them. They are now marketing that instrument as a motor test instrument.

Conclusion

Based on the experience and results to date, the broad conclusion emerges that the field testing of motors is likely to be quite useful. Recommendations were made to proceed with the second phase, which involved testing of about 1,100 motors. In addition to the motor testing, it was recommended that additional measurements on end-use equipment (like pumps) be made so that the total system performance and efficiency can be estimated. End-use equipment testing, when coupled with motor testing, is likely to result in much higher efficiency improvement potentials. Finally, a complete instrumentation package that includes a simultaneous shaft rpm measurement and recording facility will make this program more effective and convenient for large scale applications.

* * *

The assistance of the following is acknowledged: California Energy Commission: Diane Fellman, Michael Jaske, and Thom Kelly. Pacific Gas & Electric: East Bay Division, D. Yuen, J. Blessent, B. Hong, R. Purcell; and Department of Engineering Research, R.J. Renouf.

References

1. Cummings, Paul G., et al. "Induction Motor Efficiency Test Methods," IEEE Transactions on Industry Applications, 1A-17, (3)
2. Buschart, Richard J. "Justification for energy efficient motors," Proceedings of Third Annual Industrial Energy Conservation Technology Conference, Houston, Texas.
3. Buschart, Richard J. "Motor Efficiency," IEEE Transactions on Industry Applications, 1A-15, (5).
4. Lindhorst, Paul K. "The design and application of induction motors for efficient energy utilization," IEEE Industry Applications Society, Annual.
5. Gorzelnik, Eugene F. "Motor efficiency can mean big savings," *Electrical World.*
6. "IEEE Standard Test Procedure for Polyphase Induction Motors and Generators," IEEE Standard 112.
7. "American National Standard for Polyphase Induction Motors for Power Generating Stations," ANSI, C 50.41.
8. "Classification and Evaluation of Electric Motors and Pumps," DOE/TIC-11339.

Chapter 15

Testing Rewinds to Avoid Motor Efficiency Degradation*

With rising power costs it is no longer practical to consider the monthly power bill as a fixed base cost that cannot be controlled. Since motors account for over 76 percent of the power used by industry, they have always had an impact on operating costs.

New Efficiency Questions

Rising power costs (Figure 15-1) cause customers to ask what happens to efficiency when a motor is rewound, which can occur two to three times during a motor's lifetime. Many answers can be found, ranging from the observation that "a rewound motor is never as efficient as the

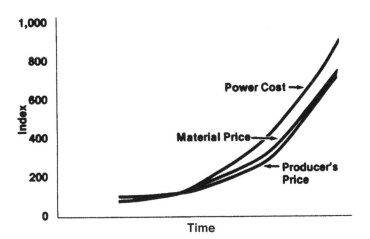

Figure 15-1. Comparison of Power vs. Other Future Costs

---

*This chapter was written by the late David Montgomery, General Electric Co., and published in the Proceedings, 8th World Energy Engineering Congress, published by the Association of Energy Engineers.

original" to "a high quality rewound motor can have a higher efficiency than the original." These answers to the same question indicate this is a complicated subject.

Motor Losses and Profits

Although motors are more efficient than any of the equipment they drive, their operating cost is of vital importance, because they account for a significant amount of the power used by industry. The total annual operating cost for a single 50 hp motor is $43,940, based on continuous operation at $.12/kWh power rates and 89.2 percent efficiency (Figure 15-2). Useful work represents $39,210 of this total and motor losses, $4,730. It is essential that these losses not be allowed to increase, since that will reduce profits year after year and a typical industrial plant may have hundreds of motors.

Increased energy costs have highlighted the problem of motor damage repair. When power rates were low, the main emphasis was on rapid unit repair and a quick return to service. There was little concern if the motor drew more current and the watt-hour meter ran faster. Rising power costs can no longer be completely passed on to customers and must therefore come out of profits. The problem of low quality rewinds can no longer be ignored. There is an additional concern for increases in

Operating Cost = Useful Work + Motor Losses

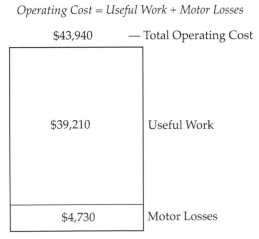

Rating: 50 hp 3600 rpm Dripproof
Continuous Operation at $0.12/kWh, 89% Efficiency

Figure 15-2. Annual Operating Cost for a Single Motor

maintenance costs and lower productivity due to the reduction in motor reliability. It is not a wise business decision to attempt to save money with a low cost, low quality motor rewind.

Motor Loss Categories

Motor losses can be divided into four categories (Figure 15-3). The largest is stator and rotor I^2R losses, which are almost 50 percent of total losses for a 50 hp drip-proof 3,600 rpm motor at full load. Other losses are stray load, friction and windage, and core losses.

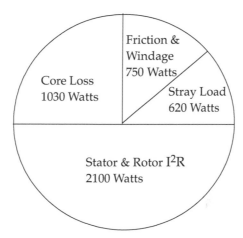

Figure 15-3. Losses for a 50 HP 3,600 RPM DP Motor

Friction and windage losses, which are the result of friction in the bearings and air friction against the rotating rotor and ventilation air paths, do not vary with load. Core loss is the sum of the eddy current and hysteresis loss associated with energizing the motor's magnetic circuit. These losses normally do not vary with load.

Stray load losses are a result of leakage (harmonic) flux induced by load current. They vary as the square of the load current. The losses associated with a stator winding are primarily I^2R losses caused by current flowing through a resistance, which in this case is the stator winding. There are also I^2R losses in the rotor bars and end rings.

Total motor losses vary with load as shown in Figure 15-4. As the motor load increases, the motor will draw more current, which increases both the I^2R and stray load losses. As indicated, core loss and friction and windage losses do not vary with load.

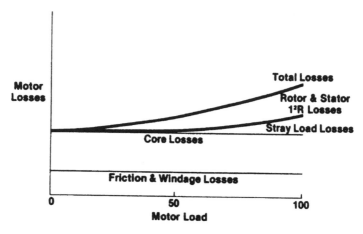

Figure 15-4. Motor Losses vs. Load

Thus, while core losses can be approximately 25 percent of the total losses at full load for the 50 hp motor, they represent ever higher percentages at reduced loads; for example, 30 percent at 75 percent load, 40 percent at 50 percent load, 50 percent at 25 percent load and 60 percent at no load. This is significant because most motors do not operate at nameplate or 100 percent load. The majority of motors in this country operate somewhere between 50 percent and 100 percent for continuous duty applications. For duty cycle applications, they can operate at all loads, including extended periods of time at no load.

Motor Rewinding Impact

Since all types of motor losses can be adversely affected during a careless or low quality repair, each problem will be reviewed individually in some detail.

Stator Losses

If a repair facility does not have the correct wire size in stock and uses a smaller diameter wire for the rewind, the stator I^2R losses will increase.

It may also be tempting to make it easier to rewind a motor by reducing the number of turns. This attempt to save labor can be very costly to the motor user in motor efficiency, operating costs, and reliable operation. A reduction in turns will lower the winding resistance and the stator I^2R losses. However, this change from the original design will increase the magnetic field. If the original design had a low-flux density such as with a U-frame motor, the increase in the core losses may be offset by the reduction in the I^2R losses. With a high-flux density design such as a T-frame motor, the increase in core losses that result from a reduction in the number of turns will not be offset and efficiency will be reduced.

A change in the number of turns will move the load point where a motor's peak efficiency occurs. For most integral hp motors, the peak efficiency occurs below full load, and most motors are used on loads that are less than the nameplate rating. Decreasing the turns will move the peak efficiency toward higher loads, which could reduce the efficiency for the actual operating point.

Changing the number of turns must further be viewed with caution because it will also increase starting current and starting and maximum torque. A change from 10 to 9 turns will increase starting current 23 percent, which may cause problems in the distribution or protection system or on the driven machine.

Windage and Friction Losses

Losses not normally affected by a rewind are windage and friction losses. However, these losses can be increased by a change in bearings or grease, by overgreasing, fan substitution, or by assembly procedures.

Rotor Losses

During a rewind, if the stator turns are reduced, the flux density increases, which reduces the rotor I^2R losses and reduces the motor power factor, an undesirable result. A lower power factor increases the losses in the distribution system which will increase power costs. If the utility charges a power factor penalty and/or a demand charge based on kVA usage, there will be an additional increase in the power bill. For this reason and those previously covered under stator losses, it is most important that the stator turns are maintained at the original design level.

Stray Load Losses

Stray load losses include additional fundamental and high frequency losses in the iron, the strand and circulating current losses in the stator winding, and harmonic losses in the rotor conductors under load. These losses are proportional to the rotor current squared. Damaged stator or rotor cores and frames or end shields that affect air gap symmetry can increase stray load losses. The stray load losses will be increased if the rotor is turned to eliminate a rubbing problem and a dull tool is used that smears the iron.

Core Losses

If in repairing a motor, the rotor must be turned, the air gap will increase. This change will increase the no load current, which results in a lower power factor and higher I^2R losses. If an end-shield repair or replacement is required, and the rotor and stator are longitudinally misaligned, there will be a reduced air gap area across which power can be transmitted, increasing the I^2R losses.

Eddy current losses in the iron are caused by circulating currents induced in the magnetic iron. These losses can be decreased in design and manufacture by making the core laminations thinner and by having adequate insulation between the laminations. During repair this insulation can be degraded by mishandling, burrs, or assembly pressure.

The greatest risk of core damage occurs during the stripping of the old winding in the repair process. If the insulation burnout is performed without temperature control, it is very easy to get the laminations too hot. This heat will break down the insulation between laminations, resulting in dramatically increased eddy current losses. The old winding can be safely burned out, providing oven temperatures are kept under careful control.

Costly Low Quality Rewinds

Motor rewinds can have hidden costs. Operating costs will go up if the efficiency is adversely affected. The additional losses associated with the lower efficiency can cause premature failures, resulting in additional maintenance cost, and lost production due to increased motor downtime.

A 50 hp, 3600 rpm, drip-proof motor that has been previously rewound can experience an increase in core losses. Four realistic possibilities where core losses have increases of 50, 100, 150 and 200 percent are shown in Figure 15-5. If the core losses increase by these percentages, the additional watts in the motor will increase from 515 to more than 2,000 watts. That is equivalent to adding a good-sized space heater inside the motor. These additional losses will substantially increase operating costs. If the motor operates continuously, as many motors do, with electricity at 12¢/kWh, the annual increase in operating cost would be more than $2,000. Figure 15-5 shows the percent of the rewind cost this increase represents.

Assume that a rewind facility rewinds a 50 hp motor at a 20 percent discount from the typical price, but in the process this increases the core loss by 100 percent. This represents an increase in losses of 1,030 watts and would greatly raise operating costs for this motor.

Core Loss Increase		Annual Operating Cost Increase
50%	515 Watts	$542
100	1030	1,080
150	1545	1,626
200	2060	2,168
50 hp 3600 rpm Dripproof Continuous Duty		

Figure 15-5. Core Loss for Low Quality Motor Rewinds

Reduced Insulation Life

Most major users have experienced the premature failure of a repaired motor, even though the original motor operated satisfactorily for years. After the motor was rewound, it failed in a very short period of time. Such failure was often attributed to defective materials or poor workmanship. While this could have been the case, another reason for premature failure could be that the stator core was damaged during the stripping process.

	Cost as Percent of Rewind Price				
	Years				
	1	2	3	4	5
Rewind at 20% Discount	80%				
Operating Cost Increase	55%	55%	55%	55%	55%
	135%	55%	55%	55%	55%
5 Yr. Total: 355%					

Figure 15-6. Five-Year Cost of Low Quality Rewind

To see how this can happen, consider the previous example where there was a 100 percent increase in core loss. With an additional 1,000 watts, there would be a corresponding 14°C temperature rise, as shown in Figure 15-7. Based on the 1/2 life rule for every additional 10°C rise, the insulation life for this example would be only 38 percent of its designed value. As shown in Figure 15-7, one of the major causes of motor failure is insulation breakdown.

Core Loss Increase		Temp. Rise Increase	Approximate Insulation Life
50%	515 Watts	7°C	62%
100	1030	14	38
150	1545	21	24
200	2060	29	14
50 hp 3600 rpm Dripproof Continuous Duty			

Figure 15-7. Increased Core Loss Reduces Insulation Life

Shortened Grease Life

Grease life also changes with temperature (Figure 15-8). As bearing temperature increases, the grease life drops rapidly. Looking at the 50 hp motor again and assuming the core loss went up 100 percent (Figure 15-9), the temperature rise will increase 14°C, which will reduce the grease life to only 69 percent of normal. This means the motor user will have to regrease more often or the motor will experience bearing failure, another major cause for motor downtime.

Maintenance Cost

Low quality motor rewinds can definitely increase maintenance costs due to more frequent greasing, premature failures, and additional in-and-out expenses. These in-and-out expenses can be considerable, because when a motor fails it is necessary to disconnect the motor from the driven

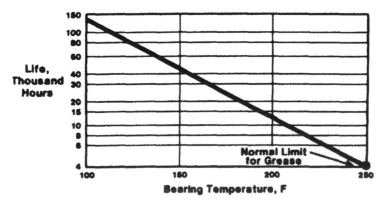

Figure 15-8. Higher Bearing Temperature Reduces Grease Life

Core Loss Increase		Temp. Rise Increase	Approximate Grease Life
50%	515 Watts	7°C	85%
100	1030	14	69
150	1545	21	58
200	2060	29	46
50 hp 3600 rpm Dripproof Continuous Duty			

Figure 15-9. Core Loss Increase Affects Grease Life

equipment, remove the mounting bolts, disconnect the motor electrically, and then arrange to have the motor repaired. When the repaired motor is returned, it must again be reinstalled.

Low quality motor rewinds can also reduce productivity due to increased downtime and lost production. Some companies assign a value to downtime as costing tens of thousands of dollars per hour.

Motor Rewind Myths

There is a common belief that when a motor is rewound a loss in efficiency of 1 or 2 points is frequently experienced. Another way of expressing this is to say that after rewind the motor's losses will be 20 percent higher. If this type of efficiency degradation is experienced, the increase in losses will be significant (Figure 15-10).

Efficiency	Losses	Increase
89.2% Original Design	4.5 kW	Base
88.2% One Point Loss	5.0	0.5
87.2% Two-point Loss	5.5	1.0
87.4% 20% Loss Increase	5.4	0.9
Rating: 50 hp 2600 rpm Dripproof		

Figure 15-10. Losses at Various Efficiencies

The only loss that can change is the core loss, if the motor is rewound to the same winding wire size and configuration and there is no change in the rotor, bearings, grease, fan or other parts. Figure 15-11 shows how much the core loss would have to increase to experience a 20 percent increase in total losses.

Motor Rating—DP	50 hp 3600 rpm	100 hp 3600 rpm	50 hp 1800 rpm
Original Efficiency	89.2%	92.1%	90.5%
Core Loss	1030 Watts	964 Watts	829 Watts
Total Loss	4500 Watts	6398 Watts	3897 Watts
20% Increase	900 Watts	1280 Watts	751 Watts
Core Loss—Increase 87%	130%	134%	

Figure 15-11. Twenty Percent Average Increase in Total Losses

Although this type of increase can occur, it can also be avoided by purchasing high quality rewinds. Some repair facilities guarantee that there will be no increase in core loss due to their rewinding processes.

Verifying Quality

When a motor is rewound properly, the user gets a new winding, an insulation system as good or better than the original, and new bearings. The only questionable item is the quality of the stator core iron.

Even the best rewind, with bad stator core iron, is subject to failure. Previously there was no easy way to verify this aspect of motor rewind quality. This is no longer true. There are now core loss testers that can be used to take the uncertainty out of motor rewind quality by quantitatively verifying that core losses did not increase during the repair process. The tester is used to check each motor as it is received for repair for core loss. If the core loss is high because of damage caused by a previous rewind or the current failure, this will be detected. Losses in watts/lb can be compared to predetermined standards. It is now possible to screen each core and determine whether or not the motor should be rewound. Motors that would be too costly to operate or would fail prematurely can be identified and replaced.

If the core is within acceptable limits, the old winding will be stripped out and the test repeated to verify that there has been no increase in core loss during the stripping process. An increase in the core loss can be verified by using the core loss tester before and after repair to the iron. These tests take only a few minutes and accurately reflect any change in core losses. Because of the simplicity and speed of the core loss test, it is now feasible to use this quality assurance tool on every motor.

Core Loss Problem

Before core loss testing was available, General Electric (GE) estimated that hundreds of millions of dollars were being wasted by inefficiently rewound electric motors. These estimates were based on an assumed average 20 percent increase in core losses after rewind. In analyzing the results of tests made over a one year period in GE repair facilities, it was found that on motors that had been previously rewound, the core loss was on average 32 percent higher than normal. This variance ranged from 0 to 400 percent. The extent of this core loss increase will vary, depending on who rewound the motor and the amount of care taken to prevent efficiency degradation.

If the test data are representative, they show original estimates of hundreds of millions of dollars wasted by inefficiently rewound electric motors to be low. In fact, the original estimates were very conservative. According to the tests, this wasted energy would be 60 percent greater than originally thought.

On motors that have been rewound, if the average increase in core losses is 32 percent and the variance ranges from 0 to 400 percent, a logical question is, "What percent of motors are being damaged?" The distribution of this damage by motor hp size in the GE sample is shown in Figure 15-12. An increase ranging from 0% to 9% was found on 21 to 39 percent of the motors tested. This degree of variance presents few problems. Tests showed that 8 to 10 percent of the motors had been substantially damaged (those with over 100 percent in core loss), and unless the damage was repaired they would be candidates for premature failure. The remaining 52 to 69 percent of the motors (10 to 100 percent increase) will cost considerably more to operate, even if they don't fail prematurely.

	Motor hp Size		
% Increase	5-20 hp	25-80 hp	75-200 hp
0-9%	38%	39%	21%
10-50	47	48	58
51-100	9	4	11
101-400	8	9	10
Total	100%	100%	100%

Figure 15-12. Percent Increase in Rewound Motor Core Losses

Correcting the Damage

The statement, "Once motor stator core losses have been increased, it is impossible to correct this damage," is another motor rewind myth. GE test results over a one-year period show that core losses can be reduced to the original level or to an acceptable value in 98 to 99 percent of the cases. The repair effort becomes an economic matter. The amount of time spent repairing the iron on a small 10 hp motor can be excessive when compared to the cost of a new motor, whereas with a larger motor it is frequently possible to justify the cost of a major repair such as restacking the iron.

High Quality Rewinds

If the average motor has a core loss increase of 32 percent after rewind (with some as high as 400 percent), and a core loss test detects this increase, it is possible to repair the damage. There will be dramatic reductions in core loss with a high quality rewind. This is particularly true with the high success rate cited earlier.

Figure 15-13 shows an actual example of a motor with high core loss. It was possible to reduce the loss by 51 percent, from 1,809 to 891 watts. This reduction of 918 watts represents real savings to the motor user. Depending on the annual running time and power cost, the savings can reach close to $10k over 10 years for continuous operation at $.12/kWh. High quality rewinds have greatly added value when this kind of reduction in losses takes place.

Initial Core Loss	1809 Watts
Final Core Loss	891
Reduction	918 Watts

Figure 15-13. Added Value of High Quality Rewinds

Core Loss Testing Advantages

In summary, testing core losses on all motors provides the capability to:

• Identify motors with damaged cores that should not be repaired
• Verify effectiveness of repair efforts
• Prevent increase in operating cost
• Protect against premature failures, thereby assuring greater reliability

- Increase productivity, and
- Protect motor investment.

Alternative Test Methods

The core-loss tester reviewed earlier is not the only way to test the quality of a motor's stator core iron. Some of these alternative test methods are described below.

Load Test

Although a load test can be used to verify a motor's performance, it has serious limitations when used to determine if there has been a change in core loss during the rewind process. To detect a change, it is necessary to know either the actual total losses or core losses of the motor as originally manufactured. To be comparable, the load test must be performed on a dynomometer in accordance with NEMA Test Standard MG1-12.53a (IEEE-112, Test Method B, using segregated loss determination). This type of test can take several hours, since it is necessary to wait until the temperature rise of the motor levels off. If the test is not performed accurately, a change in core loss could go undetected. This is particularly true if the core losses are lumped together with other losses by measuring them only at full load. Another disadvantage of the load test is that it can only be performed with the motor operating. Therefore, it could not be used to screen for damaged stators in failed motors.

No Load Test

The motor no load losses consist of three elements. (1) stator I^2R losses at no load; (2) friction and windage losses; and (3) core losses including both hysteresis and eddy current losses.

The stator I^2R loss can be calculated after measuring the no load current and stator winding resistance. The friction and windage losses are determined by a more complicated process. This involves measuring the no load losses at rated and several reduced voltages. The watts are plotted vs. volts squared, and a straight line is drawn through the points and extended to zero volts. The no load losses at zero volts squared are the friction and windage losses. The core losses can then be calculated by subtracting these elements (stator I^2R and friction and windage) from the total no load losses.

Although the no load test can be used to determine a motor's core loss, it is time consuming and can only be used when a motor is in operating condition.

Ring Test

The ring test is a widely accepted method of determining hot spots in damaged stator core iron. The test consists of running full load current at a reduced voltage through a series of cable turns wrapped through the stator and around one side until hot spots occur. The most common practice is to use one's hand to detect the hot spots. This is not foolproof, since the hands of many repairmen cannot detect a temperature difference unless it is 15°F or greater. Another problem with this test is that higher core losses could go undetected if the damage occurs immediately behind the cable turns, or if the increase is uniform throughout the core.

The greatest disadvantage is that it is too time consuming. As a result, it is seldom used unless there is obvious iron damage, such as actual holes caused by a winding failure or smeared iron caused by a rubbing rotor after a bearing failure. The new core loss tester has detected many bad stator cores that looked perfect and would not have been candidates for the ring test. Another disadvantage of the ring test is that there is no quantitative measurement that can be used as a point of reference for subsequent repairs.

Although the ring test will not give a quantitative core loss measurement, it could temporarily be an acceptable alternative when used in conjunction with an infrared scanner, which will more accurately detect hot spots.

Motor Failures

As stated earlier, if the old winding is stripped out carelessly it is likely to increase the motor core losses. An analysis of GE's test results shows this has happened frequently. Considering these facts, one would expect more evidence of motor failures due to improper rewinds. Unfortunately, an increase in core loss is difficult for the motor user to identify. There are a number of clues which could call attention to this quality problem, but they are often overlooked. Two of these would be higher full load current and higher operating costs. In the 50 hp example, a 200 percent increase in core loss only increases the full load current by 5 percent. A change in voltage or load could easily account for a current change of this magnitude. This increase would show up in a higher no load current or higher no load losses, but the original values for these items are not usually known. The corresponding increase in operating costs is easily clouded by the overall increase in the user's electric bill.

Two additional clues would be motor breakdown due to insulation and bearing failure. Defective materials or poor workmanship will probably be blamed for motor rewind failures that occur within the warranty period. But a major reason more motors do not fail is that most motors do not operate at their nameplate rating. For a number of reasons, a majority of motors only operate between 50 to 80 percent of their rating for continuous duty applications and frequently below this on duty cycle applications. For these conditions, the motor current will be lower than the design value, and the reduced stator I^2R losses will frequently offset many increases in core loss. Although the damaged core iron may not result in a failure, it will increase operating cost unnecessarily.

Core Loss Testing Capability

When a concept offers valuable benefits that can be verified by testing, it receives high acceptance. This has been true for core loss testing. Motor users have recognized the value of no increase in operating cost and greater reliability. They are insisting that their motors be tested for core loss as part of the rewind process.

Motor users understand the benefits of core loss testing and want all of their motors tested, particularly the larger hp form-wound designs. Figure 15-14 shows that it is even more important to prevent any increase in core losses on larger motors, because these losses represent such large blocks of power. With the new testers available, it is now possible to test and verify the core iron quality of motors rated as large as 5,000 hp. The original testers were designed to handle the smaller hp random-wound designs. On larger ratings (300-500 hp sizes), the tester would frequently run out of capacity, particularly in those cases where the motor had badly damaged iron.

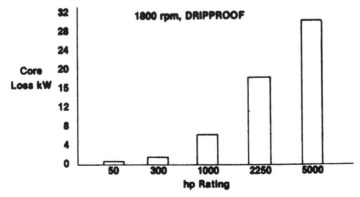

Figure 15-14. Motor Core Loss by Horsepower

Protecting Motor Investments

Motors represent a significant capital investment that must be protected. High rewind quality can now be verified by the core loss tester. When selecting a rewind vendor it is important to determine how the old winding will be stripped and what steps will be taken to ensure no increase in core loss. Before a problem can be solved, it must first be recognized and understood. Since the core loss tester can now be used to verify high quality, it should be requested in any rewind specification. The specification should also require the replacement winding to have the same wire size and number of turns as the original winding.

If the major criterion for selecting a rewind facility is price, quality may be sacrificed. This can be very costly over a period of time. To make a profit and stay in business, a company's costs must be in line with prices. There are many ways to reduce motor rewind costs, but unfortunately most are at the expense of quality in both operating cost and motor life. In today's highly competitive environment, the bottom line recommendation here is to demand high quality for all motor rewind expenditures and insist that the quality be verified by testing.

Chapter 16

Guidelines for Implementing An Energy-efficient Motor Retrofit Program*

INTRODUCTION

Stanford has recently completed a group retrofit of 73 HVAC motors with new energy-efficient models. This chapter describes the project's implementation through each of the following phases: initial survey, motor efficiency testing, payback analysis, and installation. Detailed procedures for testing motors are included. A simplified method that eliminates much of the required testing is also proposed. Initial trials indicate that the accuracy of the simplified method compares quite favorably to the complete testing procedure.

Stanford University is an international, prominent major research and educational center. Perhaps best known for its close ties with the electronics industries of the adjacent "Silicon Valley" area, Stanford has also developed highly acclaimed programs in areas such as computer science, medicine, and business.

The continuing development of academic programs on campus requires a matching effort to maintain and expand campus facilities. Energy management has been an important component of facilities design and maintenance. Lighting projects with quick paybacks were initially the principal focus. Later, few opportunities for simple measures such as delamping and lamp replacement remained. More complicated projects were then undertaken. Most significant of these projects were retrofits of building mechanical and control systems.

*Presented at 10th World Energy Engineering Congress by K. Wilke and T. Ikuenobe, sponsored by Association of Energy Engineers, Atlanta, GA.

Then began a study of the feasibility of retrofitting motors serving HVAC systems with new energy-efficient models. Several questions were addressed:

1.   Should motors be replaced before failure?

2.   How can the payback be determined?

3.   Will there be any adverse effect on the HVAC system?

After reviewing standard and high efficiency motor performance data, along with current energy prices, we quickly saw that group replacement *before* failure was an attractive option. As shown in a following section, extensive field testing was needed to make accurate payback calculations. Since energy-efficient motors are higher quality versions of standard motors, no adverse affects on the HVAC system are involved. In fact, through field testing we were able to correct cases where the motor size was poorly matched to the actual load.

Though most businesses will not invest in projects with paybacks greater than two or three years, Stanford (and most other similar institutions) looks more to the long term and accepts paybacks of five years or even more, especially if there are other benefits. For this project, motors were replaced with an installed payback of five years or less. However, many cases had paybacks of less than two years; the average payback of the whole project (73 motors) was three years. This suggests that selective retrofit with energy-efficient motors is an attractive option for many companies.

Carrying out motor retrofits by group replacement offers several advantages: easier coordination of payback analysis and installation, price discounts with group orders, and the ability to develop a "composite payback." Energy managers who are restricted to a certain payback may be able to include motors with a longer payback by averaging paybacks from a large group of motors. One disadvantage of waiting until failure to replace motors is that energy-efficient motors are typically stocked in lesser quantities and the right size might not be available. Also, once a motor fails, it is no longer possible to check if the motor size is properly matched to the load.

Using the Stanford project as an example, this chapter discusses how to implement an energy-efficient motors program. The program is broken down into five phases:

1.   Review Motor Basics
2.   Initial Survey
3.   Field Testing
4.   Payback Analysis and Vendor Selection
5.   Installation and Follow-Up

A simplified yet accurate alternative to the full field testing procedure is also presented. This should be of great interest to energy managers who have avoided motor retrofitting in the past because of the time-consuming testing required.

It is important for the energy manager to keep careful documentation during each phase of the project. Since these phases will be spread over a year or more (for a large group retrofit project), it is also vital that good communication be maintained with other engineering or service technician groups at the facility. Fortunately, much of the analysis and record-keeping for this type of project becomes easy when using a spreadsheet package such as Excel. Instances where we used spreadsheets during our project will be discussed as each phase of the project is described.

The ability to select the installations with the best payback is the key to the whole program. The potential payback can be determined from four key factors: motor size, hours/year operation, load factor, and motor efficiency. The first two items are relatively easy to set, but the last two are more difficult. The simplest solution is to use nameplate data for the existing efficiency and assume that the motor size is matched to the load. This method has several problems. First, the motor may be so old that the efficiency is not listed with the nameplate data or is not legible. This is not uncommon with older motors. Second, it is possible that the nameplate may be missing or inaccessible.

Even when the nameplate efficiency is available, this value is usually not equal to the actual operating efficiency. Nameplate efficiencies are nominal values that assume an "average" operating point. Further error can result from assuming that the motor is operating at rated load, since the actual load factor directly affects energy consumption. In addition, motors loaded under 50% of rated load show a dramatic drop in efficiency. Finally, the efficiency can be affected by deterioration in motor insulation or other motor components over time.

For these reasons, an accurate payback calculation requires that more than nameplate data be used to estimate efficiency and load factor. Fortunately for Stanford, Pacific Gas & Electric Co. and the California

Energy Commission released a computer program to calculate these parameters just as we were beginning our survey. To use this computer program, a detailed test procedure had to be carried out for each motor site. The results were then input into the program, which calculated motor efficiency and load factor. The test procedure and program calculations are described in more detail in a later section.

MOTOR BASICS

The function of an electric motor is to convert electrical energy into mechanical energy. In a typical three-phase AC motor, current passes through the motor windings and creates a rotating magnetic field. This magnetic field in turn causes the motor shaft to rotate. In HVAC applications, the shaft energy is used to drive a pump impeller or fan wheel to move liquid or air.

The number and placement of the motor windings determine how many magnetic poles will be set up inside the motor. At a given frequency, the number of poles determines the speed of the magnetic field (known as synchronous speed). This relationship is shown below:

$$\frac{\text{Synchronous}}{\text{Speed (rpm)}} = \frac{120 \times \text{Frequency (Hz)}}{\text{\# of Poles}} \qquad \text{(Formula 16-1)}$$

HVAC induction-type motors in the U.S. typically have four poles and operate at 60 Hz, yielding a synchronous speed of 1800 rpm. However, when the motor is connected to the driven load, the actual shaft rpm is less than the synchronous speed. For the motor with an 1800 rpm synchronous speed, the shaft speed is typically 1740 to 1780 RPM. The difference between synchronous speed and shaft RPM is referred to as "slip."

Of course, not all of the electric energy supplied to the motor is converted to useful mechanical energy. The efficiency of the motor is defined as:

$$\text{Efficiency} = \frac{\text{Output Power}}{\text{Input Power}} \qquad \text{(Formula 16-2)}$$

Unfortunately, a complex procedure must be followed to make a precise measurement of horsepower output. Therefore, efficiency is often calculated "from the ground up" by identifying and quantifying motor losses.

These motor losses occur in several places. They can be divided into no-load losses and losses that are a function of motor load.

No-load Losses

Windage and friction losses are due to bearing friction and mechanical losses in the rotor windings.

Core losses result from hysteresis and eddy current losses in the steel magnetic core.

Stator losses (or I^2R losses) are resistive or ohmic losses that are a function of the stator no-load input current and the stator resistance.

Load Losses

Stator losses also include a load-associated component. Input current and stator resistance are measured under typical load and temperature conditions.

Rotor losses are current losses calculated in a similar manner to stator losses—but using the rotor current under load and rotor resistance at operating temperature.

Stray load losses result from "stray" current that circulates in the magnetic steel and windings. Also in this category are harmonic losses.

Energy-efficient motors are designed to reduce the above losses. Typical design modifications include longer stator/rotor cores, more steel and copper, and added insulation. These improvements reduce internal electrical resistance, which in turn cuts down losses. The extra materials do increase the cost of the motor, but this cost can often be offset in a short period of time by energy savings.

INITIAL SURVEY

To convince management that a group replacement with energy-efficient motors is a worthwhile investment, it is necessary to compile some data on the existing stock of motors. In general, motors that are neither very large nor very small will have the best paybacks. While the difference in efficiency between a standard and energy-efficient motor increases with decreasing motor size, the energy consumed also decreases with size; thus the actual savings are relatively small. Conversely, though large motors use more energy, energy-efficient models at this size are only a small fraction more efficient. At Stanford, we limited our initial survey to motors in the 7.5

to 75 horsepower range. We also limited our survey to motors that ran at least 2000 hours per year. Motors serving intermittent loads such as vacuum pumps and air compressors are generally not good candidates for retrofit. Depending on how many motors a facility has, it might be better to narrow this range even further, to perhaps between 15 and 30 hp. Our initial survey covered a total of 158 motors. For each of these motors we noted location, size, and service (fan or pump) and estimated the hours per year the motor operated. If available, we also noted the nameplate efficiency.

Once the survey data are gathered, a preliminary analysis can be made to get a rough idea of the potential costs and energy savings. Several assumptions need to be made at this point. To be conservative, one can assume that the efficiency of existing motors is equal to standard catalog (or nameplate) efficiency. This is a conservative assumption because the actual efficiencies of existing motors are typically less than the nameplate value. Thus if the estimated payback is even close to acceptable, the energy manager should continue with a testing program, since he knows that the paybacks will improve with better data.

There are several other field conditions to consider when making the initial survey. First, make special note of any motor that is under variable loading. With a variable load, such as in a VAV system, the motor must be sized to operate under full load conditions. Therefore, the efficiency and load factor testing must also be done at maximum load. Readings must be taken with inlet vanes or other modulating devices set on "full open." Taking readings at partial load might mistakenly suggest downsizing the motor. In cases where a motor is being controlled by a variable speed drive, be sure to account for the savings already produced by the drive. In this situation it is very unlikely that the additional expense for an energy-efficient motor can be justified.

In belt-driven systems, it is also very important to note the condition of the drives. Worn belts and pulleys can reduce the applied load to the motor, giving the impression of an underloaded motor. This could lead to the same mistake as above; a motor gets downsized and the next time the belt is replaced it becomes overloaded. To avoid this problem, simply replace worn belts or pulleys before testing. Be aware that it may be necessary to replace the pulleys again, since energy-efficient motors sometimes have smaller shaft sizes than their standard counterparts.

Another concern is motor frame size. A new energy-efficient motor will likely require a smaller frame size than an existing motor, even if the motors have the same horsepower rating. Purchasing a new frame for each motor can add significantly to the cost of the project. In some cases, however, new frames may have an additional maintenance benefit. For most new motors we installed frame adapters that were much less costly than an entire new frame.

Finally, note the size and condition of the electrical components that are in the motor circuit, in particular the starter and thermal overload elements (also known as "heaters"). Make sure that the size of these components is correct for the new motor. If a smaller size motor is being installed, replacement of the heaters will probably be required. If a larger motor is being installed, it may even be necessary to replace the starter. When replacing a motor with one of the same size, it is always a good idea to check for compatibility with the overload devices.

Before beginning any testing, make sure that all the survey data are as complete and well organized as possible. The best way to enter and keep track of these data is on a computer spreadsheet. This original spreadsheet will serve as a master listing that can be added to and manipulated through various stages of the project. The user can sort the information according to any variable and then rearrange the spreadsheet to make whatever analysis is necessary (while maintaining an untouched version of the master for future changes).

FIELD TESTING

IEEE Standard Number 112 is the U.S. standard that defines efficiency testing procedures for polyphase induction motors. Of the five alternative procedures outlined, we selected Method E, which involves only input and resistance measurements. Other methods require more complex testing that we decided was not necessary. In addition, the stray-load loss (one of the five motor loss components) was assumed to be a specific percentage of motor output. This factor was built into the CEC/PG&E motor efficiency computer program that we used to analyze the data.

Field Measurements

The chosen IEEE testing method requires two sets of readings to estimate motor efficiency and load factor. One set of readings is taken while the motor is loaded, and one set is taken under no-load conditions. This allows both the load and no-load components to be calculated. For each condition, the following measurements are taken:

- Input current, voltage, and power
- Stator winding resistance
- Shaft speed (rpm)
- Motor surface temperature
- Ambient temperature

The recommended procedure for making these measurements is outlined below:

Load Test

(Let motor run for one hour or more before test.)

1. De-energize motor at breaker or disconnect switch.
2. Take stator winding resistance between each line pair.
3. Hook-up power meter.
4. Energize motor and allow to run until readings stabilize on power meter. Record kW and power factor.
5. Measure input voltage across lines A-B and across B-C.
6. Measure line currents on lines A and C.
7. Measure shaft rpm.
8. Take motor surface temperature.
9. Repeat above set of readings (1-8) to obtain an average.

No-load Test

1. De-energize the motor.
2. Uncouple the load from the motor.
3. Carry out steps 4-7 of Load Test.
4. Repeat to obtain an average.

It is possible to reduce the amount of no-load testing required by using a no-load database. This is desirable because of the time involved in uncoupling and recoupling the load for the test. The CEC/PG&E program offers a limited amount of no-load data, but these data were not well matched to our existing stock of motors. Each type and vintage of motor has different no-load properties, and it is important to use data that provide at least a good approximation. After we had tested many motors we were able to generate our own no-load data, which proved quite useful. Using these data, we were able to avoid no-load tests on most pumps (which are more difficult to uncouple than belt-driven fans) and on motors with poor access.

Test Equipment

As outlined in the above procedure, a variety of measurements must be made to complete the tests. Measurements of voltage and current are common, but kilowatt and power factor measurements are more unusual. Fortunately, our electric shop has an Esterline Angus power demand meter. This worked very well for our test. There are a variety of similar meters on the market, with a range of capabilities and prices.

The easiest way to make most of the other readings is with a digital multimeter (DMM). The critical criterion in selecting a DMM is the capability of making accurate resistance measurements in the low (0-2 ohm) range, since stator resistance lies in this range. The meter we selected was the Fluke 8012A, which has a +/-2% accuracy. The DMM was also used for current, voltage, and temperature measurements. An AC current clamp (Fluke Y8101, 0 to 150 A) and temperature probe (Fluke 80T150C, –50 to +150°C) were also obtained to get these readings.

The remaining piece of equipment we used was a stroboscope to measure shaft rpm. Initially we used an insertion-type mechanical device, but we found that the stroboscope was more accurate and easier to use in places with difficult access.

PAYBACK ANALYSIS AND VENDOR SELECTION

Payback Analysis

Once all of the test data were compiled and the load factor and efficiency calculated by the CEC/PG&E program, a detailed payback analysis was made. The CEC/PG&E program can estimate a simple payback, but we found it more useful to use our spreadsheet model instead.

Annual kWh savings for each motor were calculated using the formula below:

Where:

$$A = (HP \times L \times H \times (1/\eta_1 - 1/\eta_2) \times C) \times (0.746 \text{ kW/hp})$$ (Formula 16-3)

Where:
    A = annual savings (kWh/hr)
    HP = rated horsepower size
    LF = load factor (1 = fully loaded)

H = hours/yr operation

η_1 = existing motor efficiency (from test)

η_2 = new motor efficiency

C = cost of energy ($/kWh)

The simple payback is then calculated by dividing the annual savings into the estimated cost to purchase and install a new motor. At this stage we were also able to calculate what our rebate would be from the local utility (PG&E) rebate program. The rebate amounts typically reduced our paybacks by about 15%.

After the paybacks were all calculated, we were able to sort the spreadsheet quickly in order of increasing or decreasing payback. This allowed us to see easily which locations had the best paybacks. All motors that had an after-rebate payback of five years or less were then targeted for retrofit

In compiling the list of new motors to be purchased, the two main items to specify are size and voltage. The correct voltage for each motor should be available from the survey/test data, and the motor load factor should be used to select the appropriate size. Since the load factor indicates how closely the motor size is matched to the driven load, it will in some cases suggest downsizing a motor to make it operate at a more efficient point in its load range. Though efficiency is usually relatively constant between 60 and 100% load factor, it drops off sharply below 50 or 60%. Another consideration is simply that a smaller motor will cost less than one a size larger. When downsizing motors, however, it is generally a good idea to leave some extra capacity in case the load increases. Look to size a new motor in the 75-90% load factor range rather than right at 100%. While it is almost always a good idea to downsize a motor that is loaded less than 50%, motors in the 50 to 75% range should be carefully evaluated case-by-case.

Vendor Selection

When the number of new motors desired at each horsepower size and voltage was determined, a bid package was written up. This was again easily accomplished using the existing worksheet and invoking the sort function.

Along with the bid price, we requested that the motor manufacturers submit their guaranteed minimum efficiency for each motor size. Since our test results provided accurate data for the existing efficiency and load factor, we were able to make a precise calculation of annual savings and compare different manufacturers by simply plugging in their efficiency data. A new version of the master spreadsheet was created using the efficiency and cost data from each motor manufacturer. With these numbers, we generated 10-year and 20-year life-cycle cost analyses.

At the time of this project, one manufacturer provided motors with efficiencies higher than any other motor across the range of sizes. Although this manufacturer was not the low bidder, they were competitive, and the efficiency differences made up for the extra initial cost in just two years. After 10 and 20 years, the increased efficiencies led to significant savings. A similar kind of analysis using the most current performance data should be made for any group replacement project

INSTALLATION AND FOLLOW-UP

The 73 energy-efficient motors under the Stanford project were installed by two technicians (under the direction of a supervisor) over a period of four months. Total man-hour time was approximately 600 hours. We decided to install most of the motors after regular working hours to minimize disruption to the building occupants.

The procedure for installing energy-efficient motors is identical to the procedure for a standard motor. Very large motors (50 horsepower and above) may present special problems simply because of their size and weight. Allow extra time to transport and install these motors. Careful attention should be paid to correct wiring so that the phase sequence is not reversed, as this will cause the motor shaft to spin with the wrong rotation. Also be sure to follow closely the manufacturer's recommendations for greasing new motors.

One final consideration is the disposal of the old motors. If utility rebate money is involved, there may be restrictions on reusing the old motors. In some locations it may be possible to sell the old motors for a nominal amount to a scrap house or refurbishing shop.

When the new motors are installed, amp readings can be taken to verify that the motors are in fact drawing less current (assuming that the voltage and power factor have not changed significantly). In a couple of cases, our technicians were puzzled to find that after installing the new motors, the amperage draw actually went up! After checking over the readings several times, we realized that the problem was not in the motor but in the drive. When new pulleys and belts were installed along with the new motor, the wrong pulley size was used, causing the increase in current.

In addition to taking current readings immediately after installing each new motor, it is also a good idea to perform more complete spot testing. This should be done after the motors have been operating for several weeks or more. We decided to take one motor of each size that was installed, and we found that energy consumption was reduced (compared to the old installation) and that the new motor efficiency was equal to or greater than the guaranteed minimum.

A SIMPLIFIED APPROACH TO TESTING AND EFFICIENCY CALCULATIONS

Motor efficiency is essentially a function of energy losses. A precise calculation of efficiency requires that all motor losses be measured, which means that both load and no-load tests must be made. Unfortunately, the no-load test can be a very time-consuming procedure. The use of a no-load database (as earlier described) can help save time, but testing is often still required (as in our project) to supply the no-load data. Any method or procedure that further reduces the time required to determine load factor and efficiency is a great benefit to the energy manager. Though it is tempting to devise a "quick and easy" formula, the accuracy of such a formula must be empirically tested over a range of motor sizes and operating conditions. Since the field testing phase of the Stanford project generated this kind of data, we had an excellent basis for developing a simplified approach. Our objective, then, was to develop a set of relationships to calculate load factor and efficiency, using measured values for only the applied voltage, line current, and power factor. This method eliminates the need for no-load testing, as well as the load tests for motor speed and stator winding resistance. If nameplate rpm is not available, an rpm measurement is necessary to estimate slip.

In order to understand our model, it is helpful to review some basic efficiency and load factor relationships. It was stated earlier that efficiency can be expressed as:

$$\text{Efficiency} = \frac{\text{Output Power}}{\text{Input Power}}$$

or,

$$\text{Efficiency} = \frac{\text{Input Watts} - \text{Losses}}{\text{Input Watts}} \qquad \text{(Formula 16-4)}$$

Now the load factor (L) is simply:

$$L = \frac{\text{Operating (actual) Output Power}}{\text{Rated (nameplate) Output Power}} \qquad \text{(Formula 16-5)}$$

This expression can also take the form

$$L = \frac{\text{Input Power} \times \text{Operating Efficiency}}{\text{Rated Input Power} \times \text{Rated Efficiency}} \qquad \text{(Formula 16-6)}$$

The motor operating efficiency is primarily a function of motor size and load factor. Generally, efficiency increases with load factor, peaking at approximately 75% load factor. Full-load efficiency increases with rated output (motor size). Motor speed can also influence efficiency. The difference between synchronous and shaft rpm (motor slip) is a measure of losses in the rotor windings. The higher the slip, the lower the efficiency (for a given motor size) and vice-versa. Slip is commonly expressed as a percentage:

$$\text{Slip} = \frac{\text{Synchronous Speed} - \text{Shaft Speed}}{\text{Shaft Speed}} \qquad \text{(Formula 16-7)}$$

Although slip is typically expressed as a percentage of synchronous speed, in calculating motor losses slip is expressed per power unit (shaft speed).

As shown in Formula 16-5, the load factor is in turn partially a function of efficiency. Consequently, a precise calculation of efficiency and load factor (based on input measurements) must be done iteratively to achieve "convergence." (This is how the CEC/PG&E model works.) In taking a simplified approach, however, we will assume (for the purpose of calculating the load factor) that the efficiency is equal to the rated efficiency. Since efficiency is relatively constant above 50% load, this method will be quite accurate for that range. The equation for load factor then becomes:

$$L = \frac{\text{Input Power}}{\text{Rated Input Power}}$$ (Formula 16-8)

The load factor can now be determined using only current, voltage, and power factor measurements. Now if we can use the input measurements and our estimate of L to determine the motor losses, we would be able to calculate the operating efficiency.

So the next step is to take a closer look at motor losses. As discussed in the section on motor basics, motor losses can be broken down into the following:

Windage and Friction Losses (W_{fw})
Core Losses (W_{fe})
Stator Losses (W_{cu})
Rotor Losses (W_r)
Stray Load Losses (W_{sl})

Note that the last two are load-associated losses and the first two are essentially independent of motor load. The stator losses have both a load and no-load component.

In order to build our model, we will need to make some assumptions. In addition to the above motor losses, these assumptions will include expressions for rated horsepower (HP), input current (I_i), rated current (I_f), input power (P), load factor (L) and slip (S). The key assumptions are:

1. $W_{sl}^* = 1.2\%$ of Rated Watts Output $= 8.952 \times$ HP
2. $W_{sl} = W_{sl}^* \times L^2$
3. $W_{fw} = 1.2\%$ of Rated Watts Output $= 8.952 \times$ HP
    (W_{fw} is independent of load)
4. $W_{cu}^* = (3 + [40/\text{HP}]) \times W_{sl}^*$
5. $W_{cu} = W_{cu}^* \times (I_i/I_f)2$
6. $W_{fe} = 0.6 \times W_{cu}^*$
    (W_{fe} is independent of load)
7. $W_r = (P-[W_{cu} + W_{fe}]) \times S$
*Designates "at full load"

Assumption 1 is suggested in IEEE Standard 112 testing method E, while the others are based on results from our tests, as well as from other literature.

With a little algebra, we can express the sum of the motor losses (W_{loss}) entirely as a function of the available parameters (HP, P, L, S, I_i, and I_f). Summing the above expressions:

$$W_{loss} = (W_{fw} + W_{sl} + W_{cu} + W_{fe} + W_r)$$ (Formula 16-9)

Rearranging and then substituting terms,

$$W_{loss} = (W_{fw} [1 + L^2) + W_{cu} + W_{fe} + (P - [W_{cu} + W_{fe} \times S)$$
$$W_{loss} = (W_{fw} [1 + L^2] + (W_{cu} + W_{fe}) (1 - S) + (P \times S)$$
$$W_{loss} = [8.952 \text{ HP } (1 + L^2) + [(3 + 40/\text{HP}) \times (8.952 \text{ HP})] \times$$
$$[(I_i/I_f)^2 + 0.6) \times (1 - S) + (P \times S)]$$

The result is a somewhat unwieldy but useful equation. While this would be quite tedious to calculate by hand each time, it only has to be entered once into a spreadsheet and can be quickly copied for each set of data.

The efficiency and load factor values generated by this simplified method compare quite well to the values obtained from the full testing and analysis using the PG&E/CEC computer program. The composite payback predicted by the simplified method is virtually identical to the payback calculated with the PG&E/CEC computer program. Individual paybacks generally fall within one year of the payback calculated by the computer program, though a few cases showed a wider discrepancy. The simplified method is most accurate for motors between 10 and 50 horsepower. A more precise equation could be devised to cover a broader range, but we wanted to keep the basic equation as simple as possible.

To summarize the steps involved in the simplified method:

1. Measure Motor Input Current, Voltage, and Power Factor
2. Calculate Input Power
3. Record Rated Current, Horsepower, and rpm
4. Calculate Slip (Formula 16-7)
5. Calculate Load Factor (Formula 16-8)
6. Calculate Losses (Formula 16-10)
7. Calculate Efficiency (Formula 16-4)

Now the annual savings and payback analyses can be completed as described previously.

As mentioned, our model is based on the data we accumulated from a test group of 50 motors. Since age and type of motors vary from facility to facility, the most reliable way to calculate efficiency is through complete testing. In cases where time is a major concern, however, the simplified method can be a great aid. We welcome any comments on applying this model or further data testing its accuracy.

As a final note, the importance of operating hours in determining the payback of energy-efficient motor replacements should be emphasized. Though we have concentrated on the need to make accurate efficiency and load factor estimates, motor operating hours is typically a more influential parameter. As a general rule, sites with very low operating hours seldom show a favorable payback, while those operating year round almost always do, irrespective of efficiency and load factor.

References
1. Lobodovsky, K.K., et al., "Field Measurements And Determination of Electric Motor Efficiency," *Proceedings of The 6th World Energy Engineering Conference*, Atlanta, Georgia.
2. *IEEE Standard 112*, "Test Procedures for Polyphase Induction Motors and Generators."
3. *NEMA Standard MG-10*, "Energy Guide for Selection and Use of Polyphase Motors."
4. "Cost Analysis of Upgrading with Energy-Efficient Motors," *Specifying Engineer*.
5. Brown, T. and Cadick, J., "Electric Motors are the Basic CPI Prime Movers," *Chemical Engineering*.

APPENDIX A

Efficiency Comparisons of Several Motors*

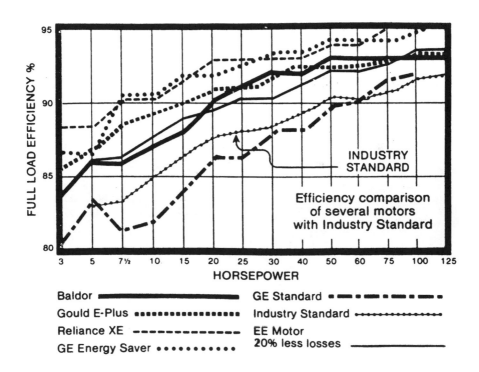

Baldor	━━━━━━━	GE Standard	●━●━●━●━●
Gould E-Plus	●●●●●●●●●●	Industry Standard	··········
Reliance XE	─ ─ ─ ─ ─	EE Motor	
GE Energy Saver	●·●·●·●·●	20% less losses	━━━━━━━

*Published by special permission: *Energy Engineering, Vol. 86, No. 3,* "Electric Motors Premium vs. Standard," K.K. Lobodovsky.

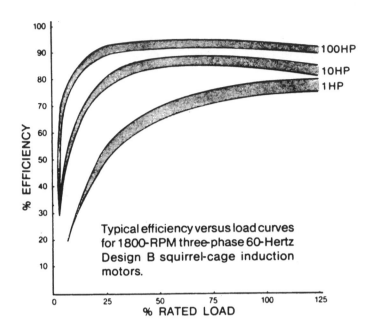

Typical efficiency versus load curves for 1800-RPM three-phase 60-Hertz Design B squirrel-cage induction motors.

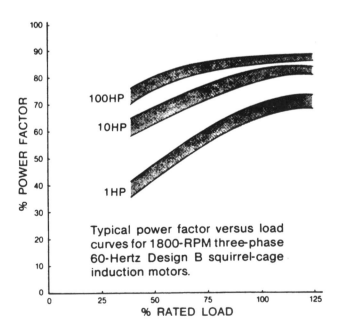

Typical power factor versus load curves for 1800-RPM three-phase 60-Hertz Design B squirrel-cage induction motors.

Savings Calculation

To determine the savings realized from the difference in operating cost between the high efficiency (premium) motor and standard efficiency (standard) motor, use the following equation

$$S = 0.746 \times hp \times C \times N \times \left[\frac{100}{E_B} - \frac{100}{E_A}\right]$$

S = Savings/yr
C = Energy Cost \$/kWh
N = Hrs/yr Running Time
E_A = Premium Motor Efficiency
E_B = Standard Motor Efficiency
hp = Motor Size

Given: $C = \$0.12$; $N = 4000$; $E_A = 95$; $E_B = 85$; $hp = 15$

$$S = 0.746 \times 15 \times .12 \times 4000 \times \left[\frac{100}{85} - \frac{100}{95}\right]$$

$S = \$665$ per year

$$\text{Simple Payback in Years} = \frac{\text{Price Premium-Price Standard Motor}}{\text{Total \$ Saved Per Year}}$$

Evaluating Efficiency Using Life Cycle Costs:

$$LCS = .746 \times hp \times C \times N \times n \left[\frac{100}{E_B} - \frac{100}{E_A}\right]$$

LCS = Life Cycle Savings (Life of Motor)
C = Average Energy Cost \$/kWh
N = Running Time Hrs/Yr
n = Number of Yrs of Operation or Period of Evaluation
E_A = Premium Motor Efficiency
E_B = Standard Motor Efficiency
hp = Motor Size

Given: 15 hp Motors $E_A = 95$ $E_B = 85$ $n = 7$ Yrs
N = 4000, C = \$0.12

$$LCS = 746 \times 15 \times .12 \times 4000 \times 7 \left[\frac{100}{85} - \frac{100}{95}\right]$$

$LCS = \$4656$

References

1. McPartland J.F., *Handbook of Practical Electrical Design*, McGraw-Hill Book Company, Third Edition, 1999.
2. Smith R.J., *Circuits Devices and Systems*, John Wiley & Sons, Inc., 5th Edition.
3. *Industrial Lighting Handbook*, National Lighting Bureau, Washington, D.C.
4. Traister, J.E., *Practical Lighting Applications for Building Construction*, Van Nostrand Reinhold Co., N.Y., N.Y.
5. Sorcar, P.C., *Energy Saving Lighting Systems*, Van Nostrand Reinhold Co., N.Y., N.Y.
6. Illuminating Engineering Society Lighting Handbook, Illuminating Engineering Society, N.Y., N.Y., 2000.
7. Getting the Most From Your Lighting Dollar, National Lighting Bureau, Washington, D.C.
8. Energy Monitoring and Control Systems (EMCS), ARMY TM-815-2, Departments of the Army and Air Force, June, 1994.
9. Ottavianio, V.B., Energy Management, Ottavianio Technical Services Inc.
10. National Electrical Code, National Fire Protection Association, Batterymarch Park, Quincy, MA, 2008.
11. Installation and Owner's Manual, Trimax Controls Inc., Sunnyvale, CA, 2008.
12. Electrical Energy Controls, National Electrical Contractors Association, Inc., 2006.
13. Thumann, A., *Handbook of Energy Engineering*, Fairmont Press, 2008.
14. Thumann, A., *Plant Engineers & Managers Guide to Energy Conservation*, 8th Edition, Fairmont Press, 2002.
15. The Engineering Basics of Power Factor Improvement, *Specifying Engineer*.
16. A New Look at Load Shedding, A. Thumann, *Electrical Consultant*, August.
17. Improving Plant Power Factor, A., Thumann, *Electrical Consultant*.

18. An Efficient Selection of Modem Energy—Saving Light Sources Can Mean Saving of 10% to 30% Power Consumption, H.A. Anderson, *Electrical Consultant*.
19. Electric Power Distribution for Industrial Plants, The Institute of Electrical Electronic Engineers.
20. National Electrical Code: 2008.

Index

Printed and bound by CPI Group (UK) Ltd, Croydon, CR0 4YY

23/10/2024

01777670-0006